£1·95

An Invitation to Mathematics

An Invitation to Mathematics

Norman Gowar

of the Open University

Oxford New York Toronto Melbourne

OXFORD UNIVERSITY PRESS

1979

Oxford University Press, Walton Street, Oxford OX2 6DP

OXFORD LONDON GLASGOW
NEW YORK TORONTO MELBOURNE WELLINGTON
KUALA LUMPUR SINGAPORE JAKARTA HONG KONG TOKYO
DELHI BOMBAY CALCUTTA MADRAS KARACHI
IBADAN NAIROBI DAR ES SALAAM CAPE TOWN

Cartoons by Beny Kandler

British Library Cataloguing in Publication Data

Gower, Norman
An invitation to mathematics.
1. *Mathematics*–1961–
I. Title
510 *QA*37.2 79–40891

ISBN 0–19–853002–1
ISBN 0–19–853001–3 *Pbk*

Printed in Great Britain
by Thomson Litho Ltd., East Kilbride

Preface

It is a frequent disappointment and a continual puzzlement to mathematicians that so many people are put off the subject by an unfortunate experience at school or by the daunting appearance of written mathematics.

The look of a page of mathematics is indeed strange but then so is a page of French to someone unfamiliar with the language. But a page of foreign writing does not cause the reader to turn away as many do from mathematics—it provokes no more than a shrug of the shoulders. I think there are two main reasons for this difference of attitude. In the first place there is no fear that the foreign writing is describing something which could not be understood. In other words whatever the language, the purpose of normal language is well understood. But few people know what mathematics is about.

Secondly, and this is really tied in with the first point, the emphasis on the teaching of mathematics has always been, though perhaps less so now, on the language of the subject and the way that language is to be used, rather than what the language is describing and why normal language is insufficient. It is rather like spending all one's time learning French grammar without even reading or writing a passage of real French.

But that is not to say that the difficulty is easily overcome. The language of mathematics is not introduced lightly–it is an essential part of the subject–it is the way in which the subject is communicated. It is not possible to say what mathematics is about without using mathematical language.

It is possible to choose topics which require the minimum of special notation and language for their communication, but the reader does not make any real progress towards those parts of mathematics which demand more sophisticated and specialized notation. At the other extreme is the specialized textbook which requires a commitment from the reader that cannot be expected of the layman and makes it very difficult to get a sense of the subject, a feel for the way mathematicians think, an insight into what mathematicians regard as important.

The book attempts to face the challenge head on. It tackles some of the central topics of mathematics, culminating with a few steps into the world of the calculus.

At the same time, whilst trying to keep the use of symbols to a minimum, this book grasps the nettle and actually discusses the use and necessity of symbols. So don't be put off if you flick through the pages and see what appears to be mumbo-jumbo. Depending on your aptitude you may find on careful reading that you have no problem or you may indeed get stuck here and there. But you should not worry about it. Nowhere is it essential to follow every detail and the work does not get progressively more difficult from chapter to chapter. Indeed most chapters end at a more difficult level than the next chapter starts with. References back to earlier pages are also nearly always at a level of generality that does not expect fluent and full comprehension.

You will also find, if you turn the pages, some items commonly appearing in most school syllabuses. Because such terms as 'set' and 'function' arise naturally you may see why they occupy such an important place in mathematics and have laid such a strong claim to their places in the school syllabus.

The title of this book suggests it is an invitation. An invitation should give an idea of what one is being invited to and at the same time suggest it is worth making the effort to attend. Any mathematician will tell you the party is well worth going to. I hope the invitation is informative and attractive enough for you to take the step over the threshold and join in.

Finally, it is conventional to thank the publishers for the work put into a book. In this case it is far more than abiding by a convention. The staff of the Oxford University Press have gently applied the necessary pressure required when I have been in danger of missing deadlines, they have given every help and encouragement, and in all aspects of the production have shown an unobtrusive professionalism that commands the greatest respect.

Norman Gower
August 1979

Contents

1 Numbers, symbols, and shapes

Izaak Walton said in the *Compleat angler* that 'Angling may be said to be so like the mathematics, that it can never be fully learnt.'

In its history of something like 3000 years, mathematics has developed into what is arguably the richest and most creative product of the human mind. On the grounds of both its practical value and its appeal to the intellect, its contribution to our development has been enormous. Many of us leave school thinking that mathematics is a cut and dried subject, constructed many years ago by a cold, logical, predestined process to be studied subsequently by generations of unwilling or unnatural students. Nothing could be further from the truth – mathematics is neither predestined nor dead.

Today, mathematics has many growing points and only a few people have even an appreciation of the whole subject, let alone a complete grasp. Yet for all this, the real stuff of the subject is to be found in something familiar to us all – numbers and shapes.

Numbers

Mathematics has two faces, practical and abstract, each with its own appeal and each connected by the inner structure of the subject. This dual nature is well illustrated by numbers. The practical importance of numbers can scarcely be called into question and yet, as we shall see, they have a pure appeal to the intellect that has fascinated and fed some of the greatest minds through the ages.

The familiarity with which most of us are able to use our present-day number system conceals the superb feat of ingenuity that was its creation, just as, for example, we rarely give a thought to how a television actually works when we watch one.

Our number system was invented by the Sumerians and Babylonians, developed by the Hindus and introduced into Europe by the merchants of medieval times. But why should its invention have been so important? For two reasons – it offers both an economical use of symbols and ease of computation.

The history of number systems is a fascinating story in its own right, and anthropologists have used similarities in the counting systems of various tribes to make hypotheses about population drift. Many of the systems are exotic and most cunningly designed but few are suitable for anything much beyond simple counting. So what makes our system particularly effective? Why is it economical and specially suited to computation?

Its most immediately outstanding feature is that we need only the ten symbols 0, 1, 2, ..., 9 to be able to write down any number we please. Have you ever thought how this is done? Think about how the symbol 6 is used in each of the numbers

346 567 691.

In the number 346 the symbol 6 stands for six 'units'. In 567 it stands for 'sixty', or six 'tens'. And in 691 it stands for six 'hundreds'. In other words the one symbol, 6, stands for different things depending on its position – but the things it stands for are not so wildly different that each collection of symbols takes a great effort of concentration to decipher.

What is more, we can build up complicated calculations using a few basic results. For example, knowing the sums

$$\begin{array}{r} 1 \\ +4 \\ \hline 5 \end{array} \quad \text{and} \quad \begin{array}{r} 6 \\ +2 \\ \hline 8 \end{array}$$

we are able to work out the sum

$$\begin{array}{r} 61 \\ +24 \\ \hline 85. \end{array}$$

Place values: the idea of using one symbol to stand for different numbers, like the use of 3 in the numbers 362, 132, 693, for example, is an ingeniously economic use of symbols. An essential feature is the use of the symbol 0 to ensure that a symbol appears in the correct place, as in the distinction between the numbers 3, 30, 301, 31. The invention and use of this symbol is remarkably creative, yet today we use it without thinking.

As you know, but may not have been conscious of, addition sums can be worked out just as long as you know how to add any two of the numbers 0, 1, 2, ..., 9.

Similar remarks apply to subtraction, multiplication, and division. Each of the four arithmetic operations requires its own little tricks (like 'carrying'), but the basic idea is a powerful one – they all depend on adding, subtracting, multiplying, or dividing the numbers 0, 1, 2, ..., 9. Perhaps if school children realized that once the tables are learnt they are in a position to multiply together any numbers they like to dream up, they may feel a little more motivated towards the chore.

It is quite astonishing that the ancient Egyptians were able to perform the remarkable feats of engineering that they did with the difficult methods of calculation that were available to them. The Rhind Papyrus which can be seen in the British Museum is our most valuable record of their methods of calculation. They demonstrate a quite extraordinary tenacity and diligence with calculations concerned with practical problems.

The ancient Greeks were fascinated by the relation between numbers and the things around us. They discovered the relation between the length of a vibrating string and the pitch of the note it produces.

In contrast, the ancient Greeks were concerned more with the abstract, philosophical problems of life. They saw the Egyptian pyramids not just as a superb feat of engineering but also as a shape in its own right with its own beautiful properties. With the abstraction of a physical object to a collection of lines to be considered in their own right, mathematics as we know it today was born.

With numbers too, the Greeks had an intrinsic interest. Practical calculations were performed on an abacus; the philosophers concerned themselves with deeper things, not requiring complicated calculations. How much this attitude was forced on them by the inadequacies of their representation of numbers and how much was intrinsic to their civilization is hard to say, but they certainly freed arithmetic from its slavery to commerce and engineering. They revealed a deeper structure and pattern in numbers than had ever previously been dreamt of.

Patterns

Have you ever heard people talk of mathematicians being interested in patterns? It sounds either a trivial sort of occupation or something rather deep and subtle. We shall be seeing many examples of this interest and some of the different forms it takes.

With the Greeks' view of a number as something to be studied for its intrinsic value, it is not surprising that they found it appealing to classify some numbers as 'square' and some as 'triangular'.

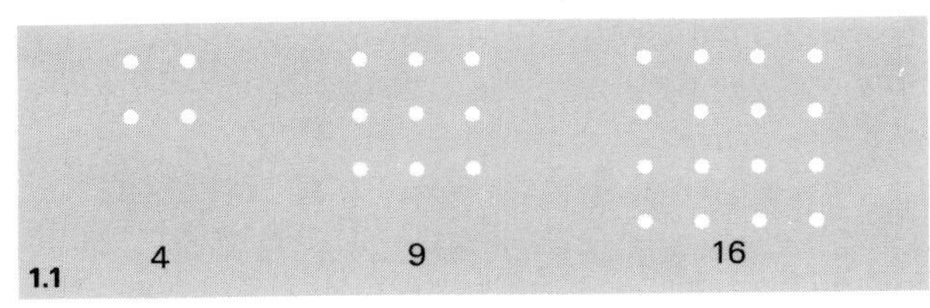

1.1

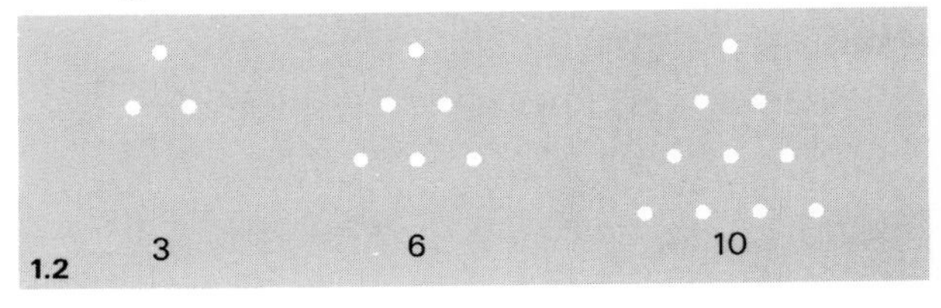

1.2

Square numbers

	1	2	3	4	5	6
1	[1]	2	3	4	5	6
2	2	[4]	6	8	10	12
3	3	6	[9]	12	15	18
4	4	8	12	[16]	20	24
5	5	10	15	20	[25]	30
6	6	12	18	24	30	[36]

Numbers on the diagonal of the multiplication square can all be represented by a square array of dots.

You can probably see that the square numbers all have a specified form: 2×2, 3×3 and so on – they are the numbers down the 'diagonal' of a multiplication table.

There is an interesting relationship between successive square numbers that can best be observed from the patterns of dots that we have used to represent them. For example, the picture that we have for 9 can be obtained from our picture for 16 by removing two borders.

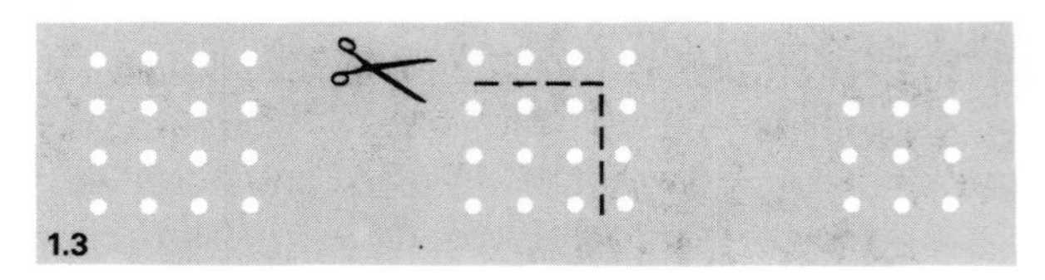
1.3

As it stands, this is hardly sensational. It tells us that $16-7 = 9$. But in mathematics there is sometimes more interest in the method by which a result is obtained than in the result itself.

Just as the removal of this border reduces the number of dots from sixteen to nine so we could regard its addition as increasing the dots from nine to sixteen.

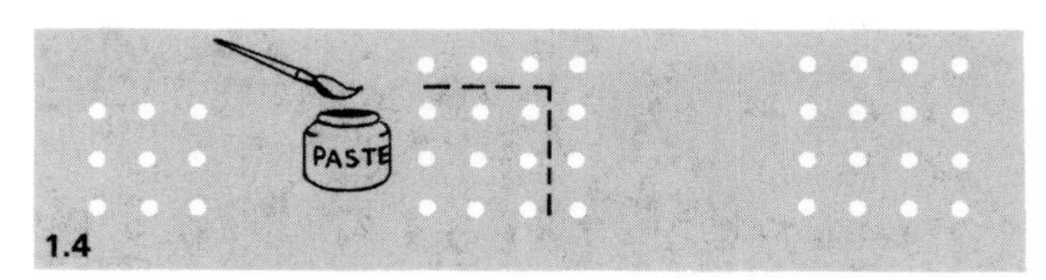

1.4

In a similar way, the addition of a border increases four spots to nine or gives an increase from one to four.

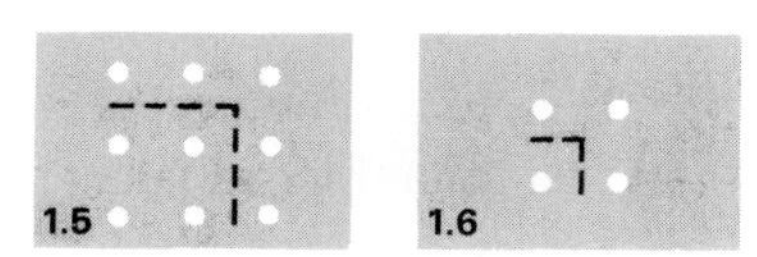
1.5 1.6

Any square number can be thought of as a sum of borders successively added to one dot. And each border contains an odd number of dots. We have the interesting result that any square number can be decomposed to a sum of odd numbers.

$$4 = 1+3$$

$$9 = 1+3+5$$

$$16 = 1+3+5+7$$

$$25 = 1+3+5+7+9$$

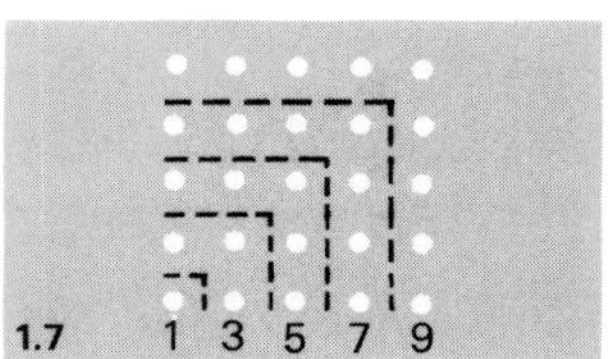

1.7

Each square number can be represented as a sum of odd numbers.
$6 \times 6 = 1+3+5+7+9+11$
$7 \times 7 = 1+3+5+7+9+11+13$

How do we know how many odd numbers we need for a given square number? For 9 we need three numbers, $1+3+5$, and 9 is 3×3. For 16 we need four numbers and 16 is 4×4. How many for 100? Remember that 100 is 10×10: we need 10 numbers. We have

$$100 = 1+3+5+7+\ldots+17+19,$$

the sum of the first ten odd numbers.

This is rather more significant than our earlier observations. Not only do we know that the number 100 *can* be expressed as a sum of odd numbers but we know just *how many*. Would you have been able to guess that the number 100 could be written in this way? Probably not, yet we know how to write *every* square number as a sum of odd numbers.

There is another way of looking at these results. Instead of thinking of $1+3+5$ as a *decomposition* of 9, we can regard 9 as the *sum* of the numbers 1, 3, and 5. That is not very special; it is easy enough to work out the sum of 1, 3, and 5. But what about the sum of 1, 3, 5, ..., 17, 19? We now know that it is 100; we could have worked it out but it would have been rather tiresome.

Can you add together the first 60 odd numbers in less than 10 seconds? If you can work out 60×60 in that time, then you should be able to beat your friends to the answer.

Now try adding the first 500 odd numbers

$$1+3+5+7+\ldots+995+997+999.$$

As a straightforward addition sum that's not so easy. But now we can work it out easily. Because it is the sum of the first 500 odd numbers, we now know that it is 500×500, which is 250 000. (Does that seem rather large for you? Try a quick check to see if it is *about* the right size. We have got 500 numbers ranging from 1 to 999 – the average size is 500 – so 500×500 is not a surprising answer.)

Patterns: mathematics contains patterns not just of the expected, geometric kind but collections of results all bearing a general similarity, one to another. If such a pattern of results can be spotted, it is often possible to guess a result of a much more general nature.

Although the results that we have discovered can be described without too much trouble, the descriptions we have employed so far have all been peculiar to the special case – the sum of the first five odd numbers is 5×5, the sum of the first ten odd numbers is 10×10 and so on. All the results have an obvious similarity of form, so strong a pattern that we might hope to be able to express it in a way that would cope with every case. The way to do this is to use a *symbol* to stand for the number of odd numbers being added. Suppose we use the symbol n. Then we can write our result as: 'the sum of the first n odd numbers is $n \times n$'.

Symbols

This last step is worth a few more words. The use of symbols is a feature of mathematics that is perhaps the most obvious of all its outward characteristics. It gives the subject that aura of inaccessibility and mystery which so often dashes the hopes of the beginner. This is a pity because, used properly, symbols add to the power and generality which gives mathematics its own unique flavour and has enabled the subject to contribute so much to our development.

Nobody could pretend that symbols are easy to get used to, but that is not to say that only a few can manage it. We all use sophisticated symbols every moment of our lives. Think of the extraordinary way in which we use the symbols of the alphabet and the symbols that stand for numbers. The use that mathematicians make of symbols is certainly not as sophisticated or subtle as ordinary language. What makes it difficult is that the ideas that mathematical language is trying to communicate are not so familiar to use as the ideas and feelings of everyday life. The collection of symbols 'toothache' communicate a complex set of emotions and experience, but ones that most of us are familiar with. Mathematical symbols are not designed to communicate the same sort of things as words, although they do sometimes do this inadvertantly, but their power can sometimes be so great as to be breathtaking even to the hardened professional.

Much of the power of symbols arises from the high density of information that can be communicated. It is quite extraordinary how the human mind can package together a set of related facts, label the package, and then use *just the label* to stand for all that it represents. When the doctor says that your child has 'measles', you do not expect a fully detailed explanation of the word, no more than the word 'aeroplane' requires with it a fully detailed set of technical drawings.

In our example of adding odd numbers, our result was 'the sum of the first n odd numbers is $n \times n$'. There, we are using the symbol n to stand for *any* positive whole number whatsoever. Our statement embraces all the special cases corresponding to the values that can be assigned to the symbol n.

To give you some practice at using symbols, let us try to find the sum of *all* the first n numbers, not just the first n *odd* numbers.

An important formula

It is a good policy when confronted with a mathematical problem to try a few special cases to give you a feel for the problem and see if they give a clue to its solution. In this example, when $n = 1$, we have the sum of just the first number, which is 1. When $n = 2$, we get the sum of the first two numbers, $1+2$, and so on. We can write down a few more, like this:

$$n = 1 \quad \text{sum} = 1$$
$$n = 2 \quad \text{sum} = 1+2 = 3$$
$$n = 3 \quad \text{sum} = 1+2+3 = 6$$
$$n = 4 \quad \text{sum} = 1+2+3+4 = 10.$$

Each of the numbers we get here, 1, 3, 6, and 10, are triangular numbers.

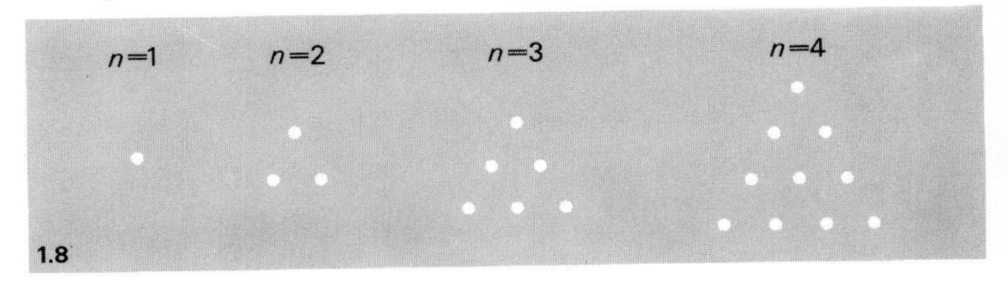

1.8

This is more than a coincidence. The bottom rows of the triangular numbers increase one at a time: the second triangle has two dots, the fifth has five, and so on. Putting this in terms of our symbol n this means that the nth triangular number is just the sum of the first n numbers – the sum that we are trying to find. So the problem we started with can be converted into a new problem. What is the nth triangular number, the number we are trying to find? Now comes the sort of jump, common to most mathematical problems, that gives some of the excitement to the subject. Let us double each triangular number.

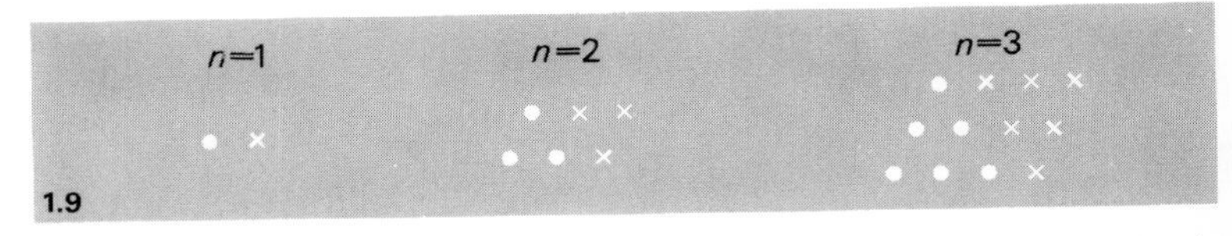

1.9

The numbers involved here (dots and crosses taken together) are much easier to calculate. For $n = 2$ we have two rows of three – so we have 6 dots and crosses. For $n = 3$ *we have three rows of four. And so on. For* $n = 20$ there would be 20 rows of 21.

How many would we have in the general case, the nth array? There would be n rows of $n+1$. Notice that now we are extending our use of symbols – we are using the symbol n not just to stand for any number in the representational sense, but as something that we can do arithmetic with: we can *manipulate* it. So whatever number n stands for, $n+1$ stands for the next one.

Brackets are used to remove ambiguity from arithmetic expressions. For example

$6\times(4+1) = 6\times5 = 30$

but

$(6\times4)+1 = 24+1 = 25$

Anything inside brackets must be calculated first.

$6-(3+1)=6-4=2$
$(6-3)+1=3+1=4$

In an array of two rows of three, there are 2×3 spots. In an array of three rows of four, there are 3×4. How many in n rows of $n+1$? It is simply $n\times(n+1)$. We put the brackets around $n+1$ to distinguish the number $n\times(n+1)$, where the 1 is added first, from the number $(n\times n)+1$, where it is added last. (So, for example, $3\times(3+1)$ is 12 and $(3\times3)+1$ is 10.)

We now have our answer to the question 'what is the nth triangular number? If the nth array has $n\times(n+1)$ dots and crosses, the nth triangular number, which is the number of dots only, must be half of this, $\dfrac{n\times(n+1)}{2}$.

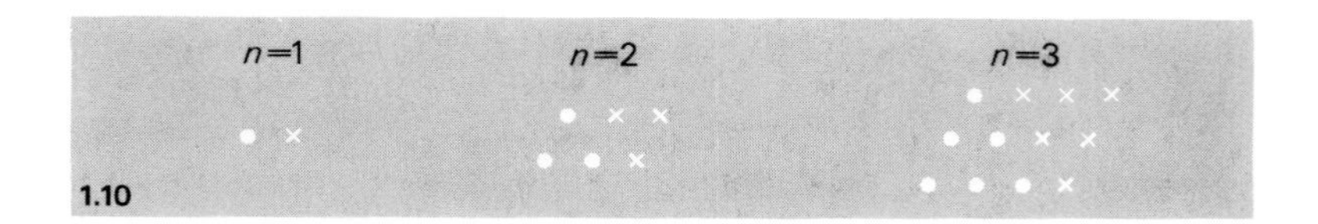

1.10

But, as we saw above, the nth triangular number is just the sum of the first n numbers, and so

$$1+2+3+4+\ldots+n \text{ is the same as } \frac{n\times(n+1)}{2}.$$

Now you can see the power obtained by introducing the symbol n: we have a general result covering all the special cases. Every result such as

$$1+2+3 = 6$$

$$1+2+3+4+5+6+7+8+9+10 = 55$$

is covered by the general formula $1+2+3+\ldots+n = \dfrac{n\times(n+1)}{2}$. Substitution of the appropriate value for n gives the same number on each side of the equality sign. Check these two yourself: in the first case $n = 3$ and in the second case $n = 10$ – work out $\dfrac{n\times(n+1)}{2}$ for each.

Karl Frederick Gauss (1771–1855) was one of the finest mathematicians of all time. The son of a bricklayer, it is said that he spotted formulae for certain arithmetic sums for himself at the age of 10. His teacher had a habit of setting the class long strings of numbers to add up to keep them occupied, all the time knowing a formula for the answer. Gauss outwitted him and all his teacher could do was to buy him a text book and announce that the boy was beyond him.

Not only does our result have the power of generality but it can be extremely labour saving. Compare the effort required to work out (even using a pocket calculator)

$$1+2+3+4+\ \dots\ +999+1000$$

with that required to calculate

$$\frac{1000 \times 1001}{2}.$$

The use of symbols in a context like 'the sum of the first n numbers ...' is similar to the use we make of them in everyday language. We are all used to the use of symbols when we say for example, 'I need a car only to get me from A to B'. The symbols A and B are being used here to stand for the end-points of all the journeys likely to be made (and, of course, to create a subtle emphasis in the statement) – just as n stands for any number.

We have now begun to see how symbols in mathematics are not just representational; they are also manipulated. This is quite a significant step; not only have we represented a whole range of possible numbers by a symbol, but we have manipulated that symbol just as though it were a number obeying all the ordinary rules of arithmetic. We can illustrate this last point by getting our latest result by an alternative method.

Using symbols creatively

Suppose we represent the number that we want, the sum of the first n numbers, by a symbol. We can use the letter S, so

$$1+2+3+\ \dots\ +n = S.$$

There is no reason why we should not add the numbers in a

different order. In particular, we can add them in exactly the reverse order and get

$$n+(n-1)+(n-2)+\ \dots\ +3+2+1 = S.$$

Now we have two expressions for S. If we add them together the answer must give us $2 \times S$. But what do we get when we add them together?

$$\begin{array}{rcccccccccc} & 1 & + & 2 & + & 3 & +\dots+ & (n-1) & + & n \\ + & n & + & (n-1) & + & (n-2) & +\dots+ & 2 & + & 1 \\ \hline \end{array}$$

$$(n+1)+(n-1+2)+(n-2+3)+\ \dots\ +(2+n-1)+(1+n).$$

Can you see that each of the new numbers in the sum on the left-hand side is just the number $n+1$?. How many of these have we got? Exactly n of them. So what does the left-hand side add up to? We have n $(n+1)$s, so we have $n \times (n+1)$ altogether.

So we have the result that twice S is just $n \times (n+1)$, so S must be half of this; so

$$S = \frac{n \times (n+1)}{2};$$

the result we had before.

The use that we have made of the symbol S here is rather more subtle than our use of n in the sense that it is introduced solely for the *derivation* of the result rather than its description. Its introduction is pure creation, its function is entirely internal and it can be discarded once its purpose is served.

So now we have a formula for the sum of the first n *numbers* and a formula for the first n *odd numbers*. What about the sum of the first n *even numbers*?

We need not tackle this problem completely afresh; we can use one of our previous results. Each even number is divisible by 2, so we can write for example

$$2+4+6+8 = 2\times 1+2\times 2+2\times 3+2\times 4.$$

This is the same as twice $1+2+3+4$.

The sum of the first n odd numbers is $n \times n$.
The sum of the first n even numbers is $n \times (n+1)$.
The sum of the first n numbers is $\frac{n \times (n+1)}{2}$.

How do we use this to get the sum of the first n even numbers? What are the first n even numbers? The first is 2. The second is 4 (which is 2×2); the third is 6 (which is 2×3); so the nth is $2 \times n$. So the sum of the first n even numbers is

$$2+4+6+8+ \dots +2 \times n = 2 \times (1+2+3+ \dots +n)$$

(Notice again the use of brackets: we have to add up all the numbers inside the brackets before multiplying by two.)

We already have a formula for $1+2+3+ \dots +n$; it is $\frac{n \times (n+1)}{2}$ so twice this is just $n \times (n+1)$. So we have

$$2+4+6+8+ \dots +2 \times n = n \times (n+1).$$

Taking stock so far

There are other similar formulas that we could look at, but our main interest is not in collecting together as many results as possible; our three results are sufficient to indicate something of much deeper interest.

The most obvious feature is the use of symbols. We have already made some remarks about this, but it is important to realize the power that they give to a mathematical argument. We have seen how they enable us to make statements of a general nature, that take on particular forms when the symbol is assigned a value. And by manipulating the symbols we can deduce new statements of a general character, as when we deduced a formula for the sum of the first n even numbers from that for the first n numbers.

Anyone who does more and more mathematics acquires a better and better facility at manipulating symbols, and converting collections of symbols from one form to another. There is nothing very special about this ability, although it is achieved only through practice. We shall try to keep technical tricks to a minimum, but you should find that your management of symbols improves as you progress through this book. In the next chapter, we shall go through a few special techniques that you will find useful.

The way in which we derived the sum of the even numbers from a previous result is also an illustration of the second general point to be made: the readiness with which mathematicians seek to exploit results that are already known. The subtle use of the result for the sum of the first n numbers enabled us to derive a new result without starting from scratch.

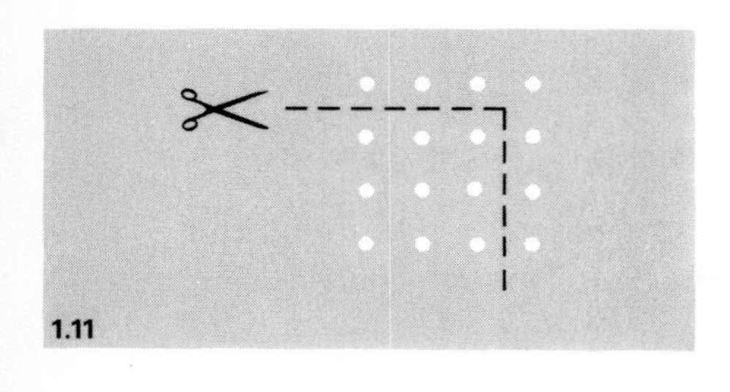
1.11

The third point to be made about our examples is that significant new information can often be obtained from an imaginative look at an apparently insignificant procedure.

We could have easily viewed the removal of some dots as giving the interesting result that two successive square numbers differ by an odd number and no more than that. As we saw, a little reflection and slight reinterpretation gave us our first result. Let us look at a similar situation, but this time involving shapes rather than numbers.

Arranging shapes

Have you ever played with one of those child's toys in which various shapes have to be fitted together to form a pattern? One of the challenges is to form four triangles and two squares into a larger square – like this.

Cut out a square with sides about 4 cm long. Cut out another square (B) from one corner. Continue the cuts to make the square A. Divide each of the rectangles that remain into two triangles.

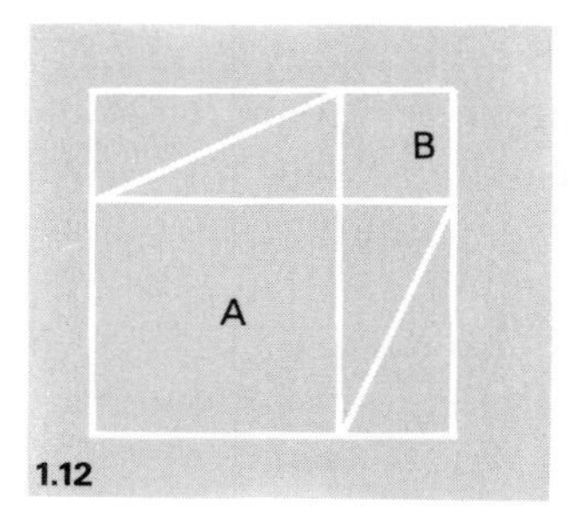

1.12

There is a considerable satisfaction in working this out for oneself, but there is still more to be gained from careful contemplation. The large square could have been built up in a different way – using the same four triangles and a different square.

Fit the triangles together like this on another piece of paper and you will be able to cut out another square, C.
This new arrangement should fit exactly into the hole you made when cutting out the very first square.

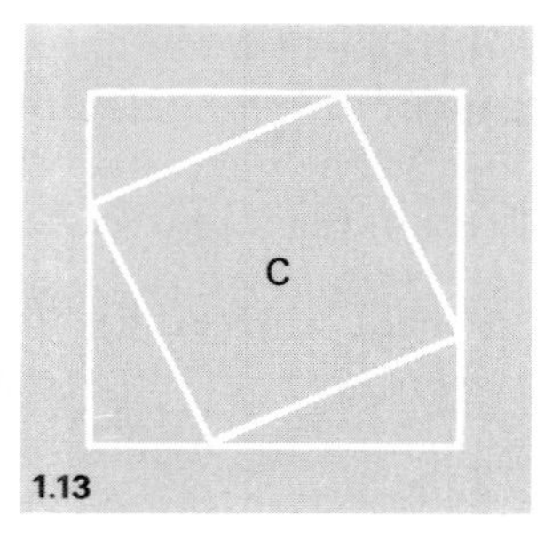

1.13

The first arrangement tells us that the area of the large square is the same as the area of the four triangles plus the area of the square A plus the area of the square B. The second tells us that it is also the same as the area of the four triangles plus the area of square C.

What does this tell us about the areas of the three squares A, B, and C?

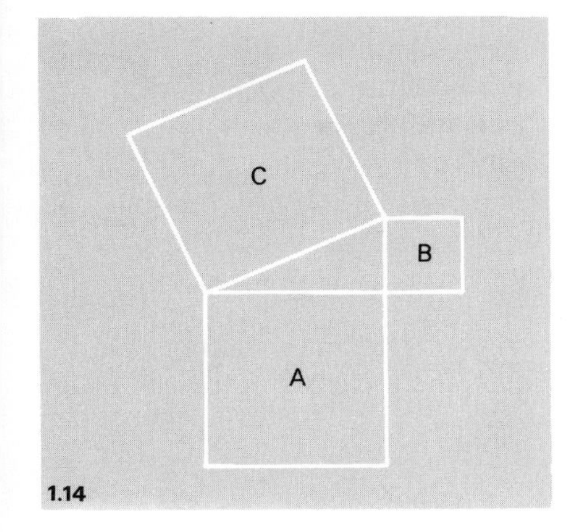

1.14

It tells us that

$$\text{Area of C} = \text{Area of A} + \text{Area of B}.$$

Now do some more rearranging.

The area of the square drawn on the longest side of the triangle is the sum of the areas of the squares drawn on the other two sides. This remarkable discovery was made by the followers of the great Greek mathematician Pythagoras. The discovery is all the more remarkable because it works whatever triangle you start with provided one of the angles is 90° (a right-angle). If you fit together four identical triangles of this type like this,

The triangles can be any shape as long as one of the angles is 90°.

1.15

then you can reconstruct a diagram of the type we started with

1.16

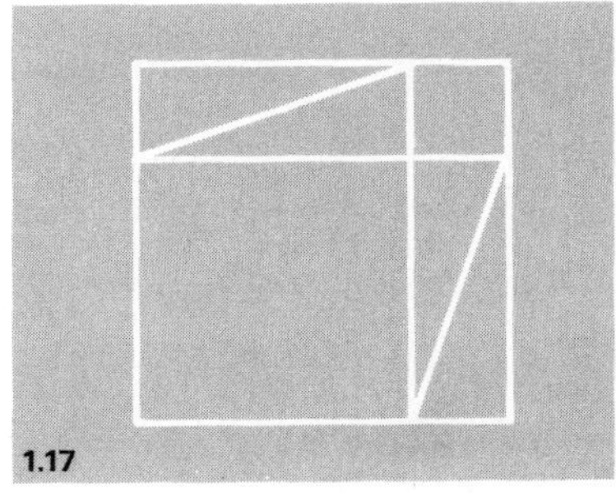
1.17

and apply the same argument.

Building walls

This property of triangles is used by builders whenever they want to set out a right-angle, when building the corner of a wall for example. The way they do it is to use the property of an especially simple triangle.

A square whose side measures 3 metres has area 9 square metres and one whose side measures 4 metres has area 16 square metres and so the sum of the areas of two such squares is 25 square metres. This is where the simplicity comes in, because 25 square metres is simply the area of a square whose side is 5 metres. (If you cannot see that this is most fortunate, try finding another triple of *whole* numbers like 3, 4, and 5 with this property. There *are* plenty of others, but they are certainly not as easy to find as you might expect.)

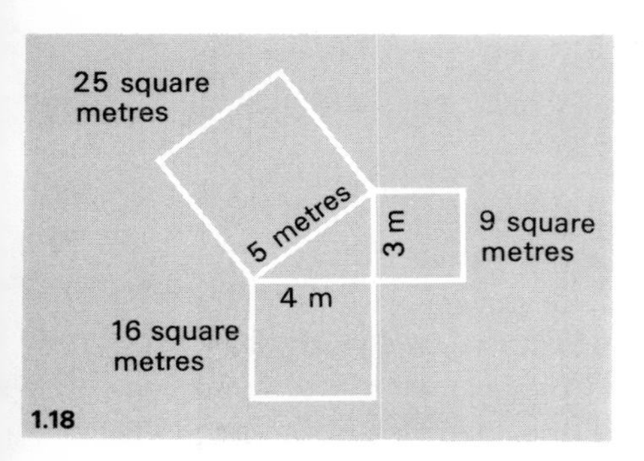

1.18

So if two sides of a triangle measure 3 metres and 4 metres respectively, then if the angle between them is a right-angle, the third side must be 5 metres.

Now how does a builder use this property? Suppose he has built a wall and wishes to turn the corner at right-angles, the situation is like this.

1.19

If he measures 4 units of length from the corner along the wall he has already built he will get to some part, A say, of the wall. Now if he builds the corner correctly, at right-angles, then the point, B, 3 units of length from the corner along the new direction will be 5 units from A. So what he can do is to mark off an arc of points 3 units from the corner and another arc of points 5 units from A.

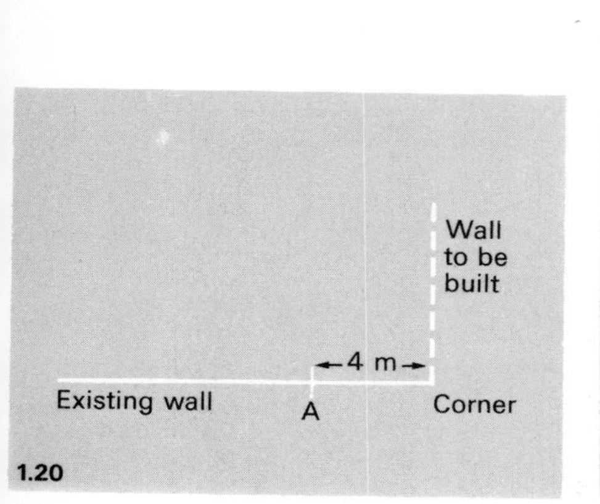

1.20

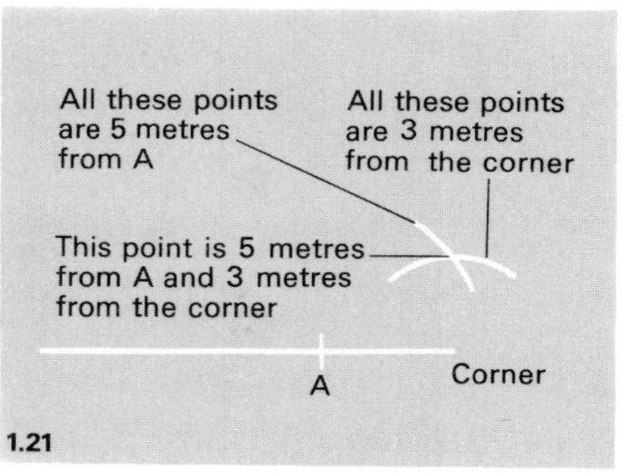

1.21

There will be a point on *both* of these arcs which will therefore be 5 units from A and 3 units from the corner.

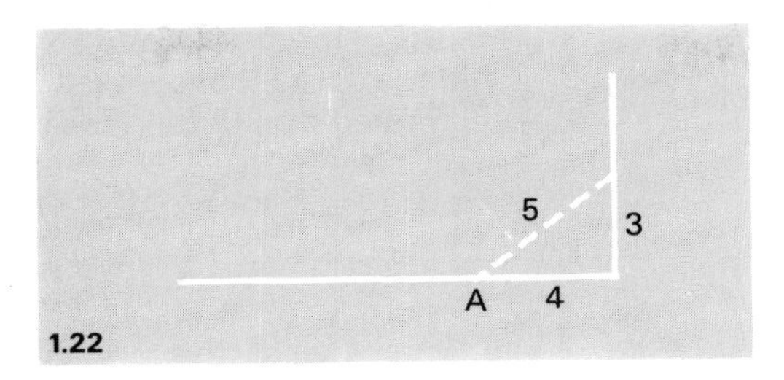

1.22

This point will give the correct line for the new wall.

Methods like this were well known to the ancient Egyptians: they were an essential part of their methods of surveying.

You may have wondered why we have talked about 'units of length'. This is simply because the principle of the method applies whatever unit of length is taken – a yard, a metre, it does not matter. The Egyptians used long lengths of knotted rope – the equal distances between the knots being taken as the unit of length. Builders today use much the same idea.

Methods of constructing 90° angles and other highly sophisticated geometric constructions were essential tools of the ancient Egyptian engineers.

A theorem

The contrast between the Greek and Egyptian attitudes to this result about right-angled triangles is a good illustration of the difference between their general attitudes to mathematics.
The Egyptians realized the practical importance of the result and their engineers and surveyors used it to the full. On the other hand, the Greeks were entranced by its aesthetic appeal. They were able to prove that the result holds for *any* triangle with a right-angle: if it has sides of lengths a, b, and c units respectively, and a is the length of the side opposite the right-angle, then

$$a \times a = b \times b + c \times c.$$

For shorthand, a multiplication of a number by itself, like $a \times a$, is usually written as a^2. The small 2 refers to the fact that the symbol a really appears twice and the expression a^2 is read as 'a squared' (referring to the fact that a^2 is the area of the square whose side has length a). So the relation we have between a, b, and c takes the form

$$a^2 = b^2 + c^2.$$

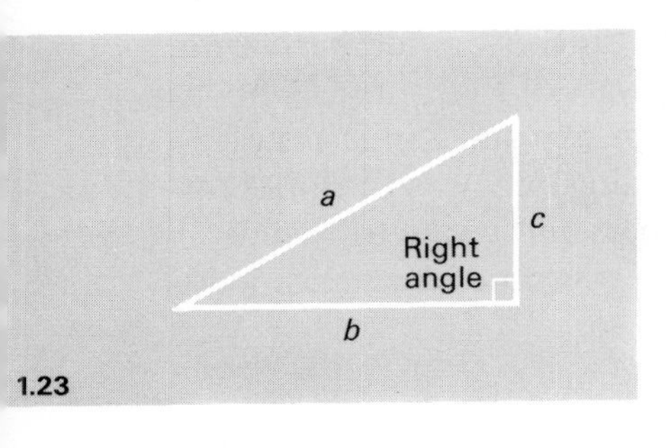

1.23

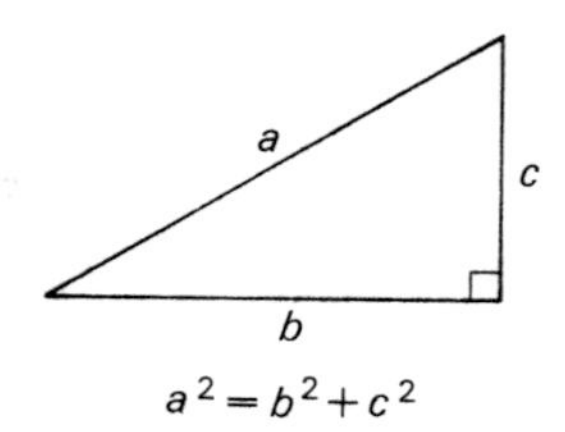

Pythagoras' theorem is one of the most widely used results in mathematics. It is said that its discovery was celebrated by the sacrifice of an ox.
It is the realization of the importance of an eternal, abstract truth that characterizes mathematics and it was the Greeks who were the first to see the significance of this way of thinking.

The Greeks were also able to prove the result the other way round: *if*

$$a^2 = b^2 + c^2$$

then the angle opposite the side of length a is a right angle.

One of the great Greek mathematicians, Archimedes, was killed by a Roman soldier whilst absorbed in the study of a geometric diagram. As was pointed out by A. N. Whitehead, this symbolized a dramatic change in world affairs: 'No Roman ever lost his life... absorbed in the contemplation of a mathematical diagram.'

These results are extremely important and perhaps the most extensively used in mathematics. It is no surprise then, that they are given a name and that the name – Pythagoras' theorem – has been known to generations of school children. Do not let the word 'theorem' put you off – it is the *result* that is important, not its name. The word *theorem* is used to refer to a result that is general enough to be of importance and yet by no means obvious – in other words requiring some proof. When a theorem needs to be referred to frequently, it is obviously convenient to give it a name. The trouble is, that once it is named, the result often seems to acquire an increased aura of difficulty.

The eternal truth described in Pythagoras' theorem was valued far more highly than the practical advantage of the particular result. This is not to say that the Greeks despised the practical applications of mathematics, but it was they who bestowed on the subject a character of rigour and abstractness that gives it its special distinctiveness.

Numbers again

We have mentioned the Greeks' interest in numbers; it was shown particularly by the followers of Pythagoras who believed that everything could be explained in terms of number. A vital part of their philosophy was the link that they established between numbers and geometry. They established a correspondence between numbers and points on a line in the following way.

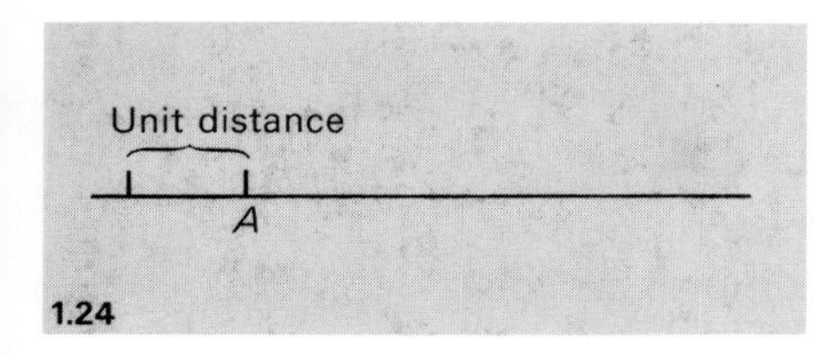

1.24

If a line is drawn and a point chosen on this line, then from this point a standard, 'unit' distance can be marked off. This point, the point A in the diagram, can represent the number 1.

Measuring the same distance off from A will produce a point which sensibly represents the number 2, and so on.

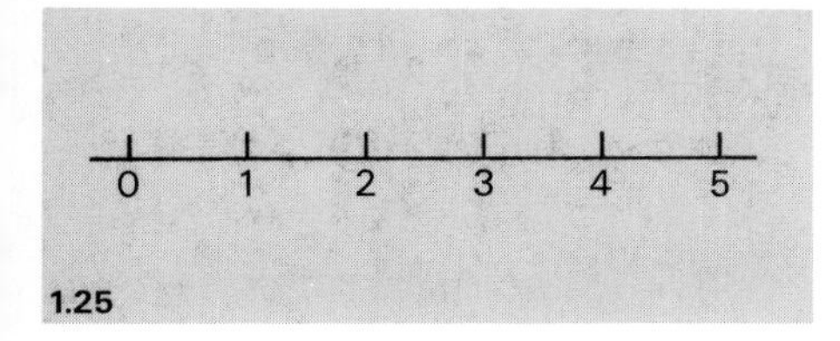

1.25

It is fairly clear how to find the point that represents $\frac{1}{2}$ – it is midway between 0 and 1. In much the same way, we can find points representing $\frac{1}{4}$, $1\frac{1}{5}$, $\frac{3}{4}$, and all the other fractions.

Any number of the form $\frac{a}{b}$, where a and b are whole numbers obviously has a point to represent it: find the point representing a, divide the portion of line between that point and 0 into b equal parts and take the point nearest to 0 to represent $\frac{a}{b}$.

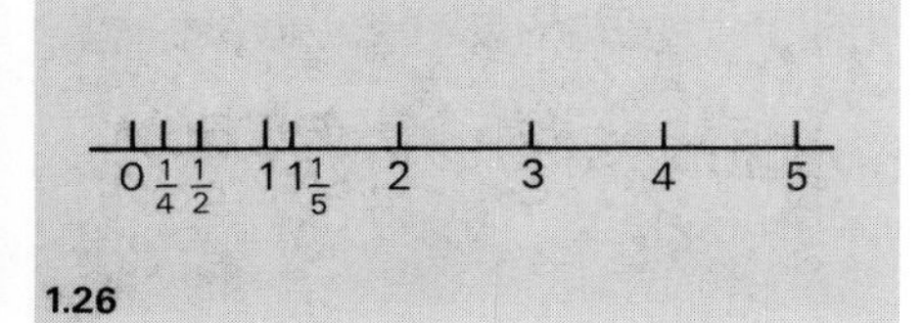

1.26

Many different fractions are all represented by the same point, for example, the point representing 18 is twice as far from 0 as is the point representing 9. So if the portion of line between 0 and the point representing 18 is divided into ten portions, the first segment will be the same length as the segment dividing the line between 0 and 9 into five portions.

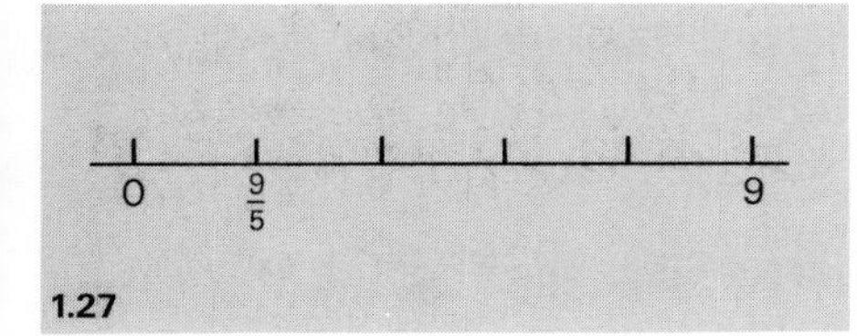

1.27

In this sense, $\frac{9}{5}$, $\frac{18}{10}$, $\frac{27}{15}$, $\frac{36}{20}$, and so on are all the 'same'. Any fraction that can be obtained from $\frac{9}{5}$ by multiplying the top number, 9, and the bottom number, 5, by the same number will be represented by the same point on the 'number line'.

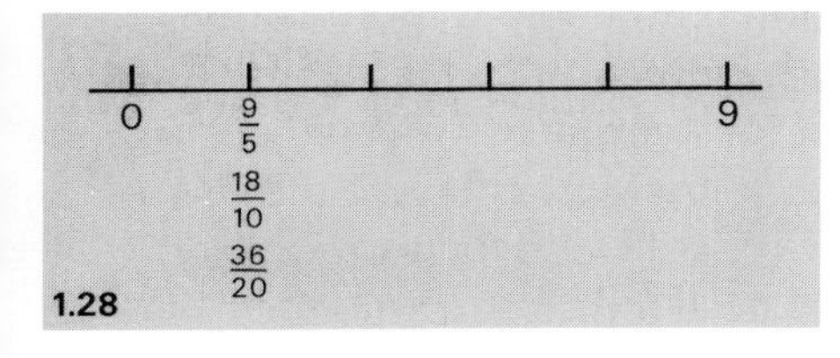

1.28

Out of all these versions of the 'same' fraction, there is an important sense in which $\frac{9}{5}$ is the simplest. In all the other versions the numbers on the top and the bottom can be divided by a common number. For example in the version $\frac{18}{10}$ there is a common number, 2, which divides into 18 and into 10. (The number 3 also divides into 18, but not into 10, 5 divides into 10 but not into 18, 2 divides into *both* 18 and 10: it is in that sense that it is a 'common' divider.)

We can express this more simply by introducing a technical term. The number 18 can be divided exactly by 3, with no remainder, and we refer to this property by saying that 3 is a factor of 18. 2 is also a factor of 18, so is 9, and so is 6. 2 is also a factor of 10 and we say that 18 and 10 have the *common factor* 2. Of all the versions of the fraction $\frac{9}{5}$, this simplest form is the only one for which the numbers on the top and the bottom do not have a common factor.

Factors: if a number divides exactly into another without leaving a remainder, it is a *factor* of that number. For example, 5 is a factor of 10 and of 15 and of 20 and so on. 5 is not a factor of 11, or 23.
The factors of 15 are 3 and 5.
The factors of 20 are 2, 4, 5, and 10. Because 5 is a factor of *both* 15 and 20 it is called a *common factor* of 15 and 20.

Rational numbers: any number which can be written as a fraction – a whole number divided by a whole number – is called a rational number.

Because of the connection between fractions and ratios of lengths, numbers of the form $\frac{a}{b}$ where a and b are whole numbers are called *rational numbers*. Because b could be the whole number 1, all the whole numbers are included in the class of rational numbers.

Irrational numbers

The celebrations that accompanied the discovery of Pythagoras' theorem soon gave way to grave doubts about the relationship between numbers and geometry.

We have seen how to construct a right-angle triangle.

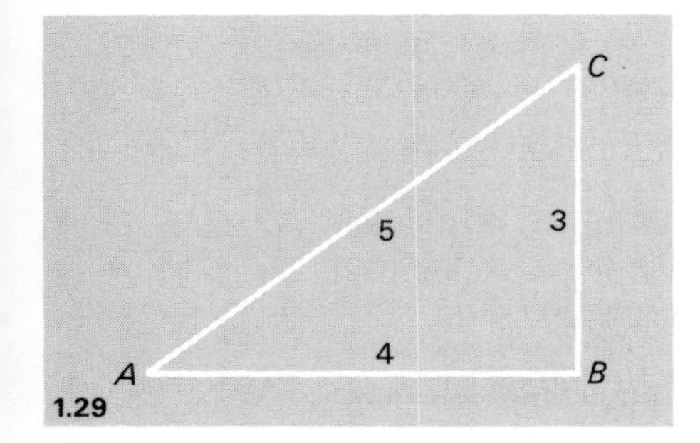

1.29

Now mark off a distance of one unit from B along each of the sides BC and BA. This gives a new triangle, PBQ, still with a right-angle at B.

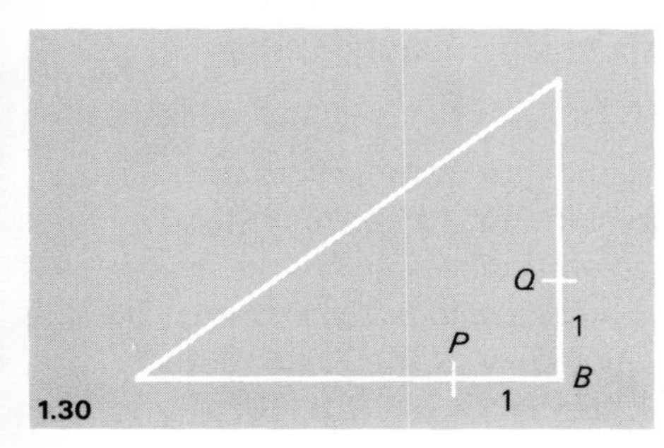

1.30

How long is the side PQ? Pythagoras' theorem tells us that if the length of PQ is x, then

$$x^2 = 1^2 + 1^2 = 2.$$

So x is a number that, when squared, gives the number 2; and the Greeks made the remarkable discovery that this is *not* a rational number. Since rational numbers were the only numbers conceived of by the Pythagoreans, this was a sensational turn of events. Think for a moment yourself – can you imagine what sort of number it could possibly be that is neither a whole number nor a fraction? Here, then was a way of constructing a length to which no number corresponded. The relationship between numbers and points on a line was shattered. Whilst it was still possible always to be able to identify a point to correspond to any given number, it was now seen not to work the other way round – it was not always possible to find a number that corresponded to a particular point.

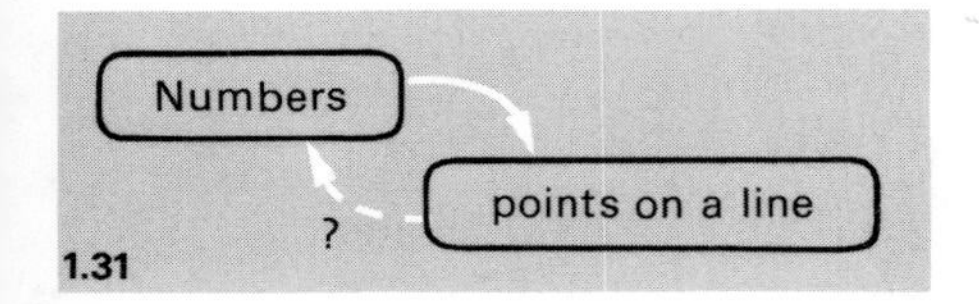

1.31

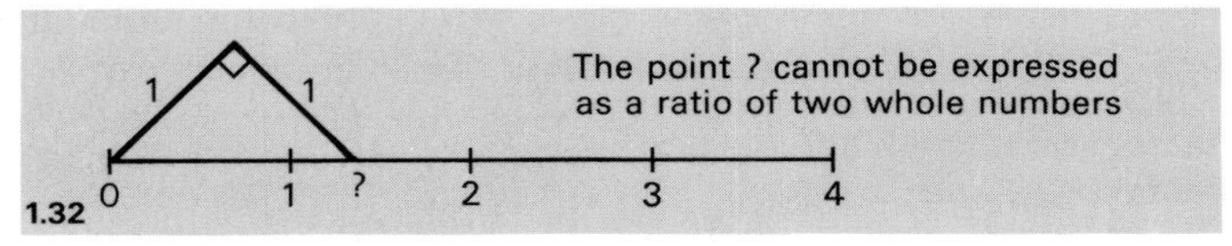

1.32

Any number that cannot be expressed as an ordinary ratio of two whole numbers is called an *irrational* number – contrasting with the term rational number.

Irrational numbers are numbers that cannot be written as the ratio of two whole numbers.

If you are not convinced of the strangeness of irrational numbers, you may like to think of the type of illustrations that are often used in schools to introduce fractions: problems like sharing out a cake. Sharing out a cake amongst four people simply means cutting the cake into four equal pieces, and this fraction of cake is written as $\frac{1}{4}$. Similarly $\frac{1}{6}$ represents the portion of cake that each of six people would get.

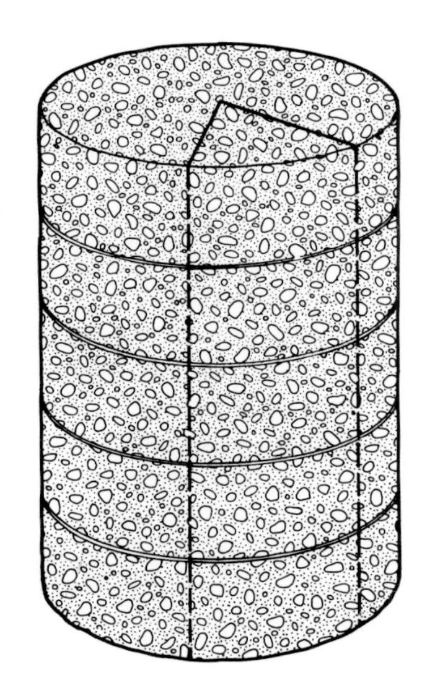

Fractions, like $\frac{5}{6}$, that do not have 1 on top can be represented in two ways. For example, $\frac{5}{6}$ is five of the six equal portions of a cake; or you can think of it as the amount you would get if you piled five cakes on top of each other and cut right down into six equal portions.

This second way of looking at fractions is a little more useful than the first because it works for fractions, like $\frac{7}{6}$, that are greater than 1. Any rational number, $\frac{a}{b}$, can be represented by piling a cakes on top of each other and then slicing the whole lot into b equal portions using vertical cuts. You might need an awful lot of cuts – think of the number $\frac{297}{296}$ – but practicalities are not really the point, after all it is not that easy to cut a cake into *precisely* three equal pieces; but the conceptual point remains. *In principle*, with enough cakes it can always be done. For an irrational number, no matter how many cakes and how many cuts, it can *never* be done, even in principle.

There is another way of looking at the strangeness of irrational numbers. For a given rational number, $\frac{a}{b}$, if you add it to itself enough times, you end up with a whole number: if you add $\frac{5}{7}$ to itself seven times you end up with a whole number, 5. For an *irrational* number, no matter how many times you add it to itself, you *never* end up with a whole number.

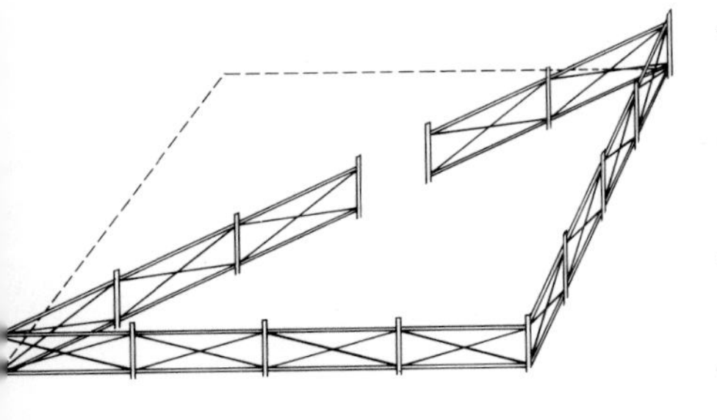

Imagine a farmer trying to divide a square field in two by putting a diagonal fence from corner to corner. He can order fencing panels of any size he wants–so he can make sure the two sides of the square fit exactly. *But, whatever* size he picks for the sides–if they fit exactly they can *never* fit the diagonal exactly with a whole number of panels.

This was what really threw the Greeks. They could mark off a number scale with lengths 1, 2, 3, 4, and so on. They could use their unit length, 1, to construct a new length, the longest side of a right-angle triangle whose other sides are each of length 1. And it turns out that no matter how many times this length is added to itself it will never total a whole number.

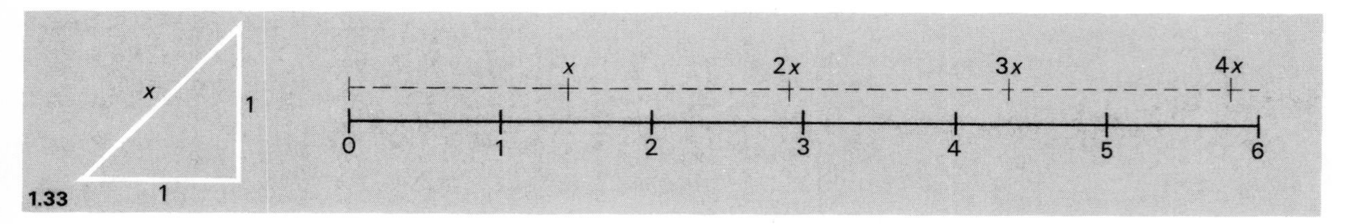

1.33

A proof

It appears, then, that if all the rational numbers are represented by points on a line, they do not exhaust all the points: there will be 'holes' in the line corresponding to the irrational numbers. It took until the latter part of the nineteenth century for mathematicians to work out how to explain the phenomenon of this new type of number in terms of the more comfortable notion of rational numbers. We shall see in Chapter 5 just how this conundrum was unravelled. We shall also be seeing how this difficulty was at the root of other difficulties that arose in the development of mathematics, particularly in the calculus.

Because of the central importance of irrational numbers, we shall go through a proof that the number x, such that $x^2 = 2$, is irrational. It will also serve as a good opportunity to see an interesting and elegant method of proof often used in mathematics.

But first, a word of advice. Mathematics is not a subject that you can read like a novel, and this applies especially to mathematical proofs. They often involve a subtle chain of logic combined with varying amounts of manipulation of symbols. It is not usually possible to get to grips with a mathematical proof at one reading and even so it is more or less essential to have a pencil and paper at hand to fill in the details of jumps in the argument that are not immediately clear. It is also a good idea to make up your own examples as you go along: substituting a few numbers in place of symbols really does help you to get grips with what is going on.

Now let us get on with the proof. It goes like this. Suppose we *can* find a pair of whole numbers a and b such that

$$\left(\frac{a}{b}\right)^2 = 2.$$

(In other words, suppose x, where $x^2 = 2$, *is* rational.)

We have seen that we can always divide out any common factors in a fraction to reduce it to its 'simplest' form. So we can always ensure that a and b do not have a common factor.

On the other hand, since

$$\left(\frac{a}{b}\right)^2 = \frac{a}{b} \times \frac{a}{b} = \frac{a^2}{b^2},$$

we must have

$$\frac{a^2}{b^2} = 2.$$

This means that

$$a^2 = 2b^2.$$

But if any number is multiplied by 2, the number you get must be an even number (try it with a few examples). So a^2 must be an even number. This means that a itself must be even. You can convince yourself of this by squaring a few odd numbers – you will find that you always get an odd number.

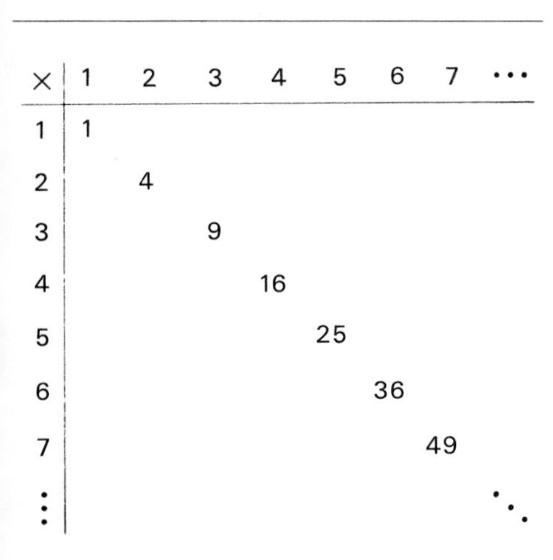

×	1	2	3	4	5	6	7	⋯
1	1							
2		4						
3			9					
4				16				
5					25			
6						36		
7							49	
⋮								⋱

The numbers on the diagonal of a multiplication square are alternately odd and even. The square of an even number is even. The square of an odd number is odd.

Now every even number can be divided exactly by 2. So this must be the case for a. Let's use the letter p to stand for $\frac{a}{2}$.

Then

$$a = 2p.$$

What will a^2 be? Just $(2p) \times (2p)$, which is $4 \times p \times p$ which we write as $4p^2$. So

$$a^2 = 4p^2.$$

But we already know that

$$a^2 = 2b^2.$$

If $4p^2$ and $2b^2$ are each equal to a^2, they must themselves be equal. So

$$4p^2 = 2b^2.$$

In other words

$$2p^2 = b^2.$$

This tells us that b^2 is an even number, so b itself must be even.

Let's collect all that together: we have proved that if

$$\left(\frac{a}{b}\right)^2 = 2,$$

then a and b must both be even. This means that they can both be divided by 2: they have a *common factor*. But we have already pointed out that we can arrange for $\frac{a}{b}$ to be in its simplest form, which means that a and b *do not have a common factor*. We therefore have a contradiction and are forced to the conclusion that our original assumption, that we can find whole numbers a and b such that $\left(\frac{a}{b}\right)^2$ is 2, must be false and that no such whole numbers can be found.

The theorem is thus proved.

Let us go through the principles of this proof again. We started off with the

Assumption: we can find two whole numbers a and b such that $\left(\frac{a}{b}\right)^2 = 2$.

From this we deduced the

Conclusions:

(i) we can always ensure that a and b have no common factor.

(ii) a and b are both even, thus having a common factor, 2.

These two deductions are clearly a

Contradiction

and cannot both be true. But they follow from our assumption and so we are forced to the conclusion that our

Assumption is false.

This method of proof – called *proof by contradiction* – is frequently used in mathematics and it is well worth spending some time on understanding its principles: read through the proof a couple more times and see if you can get to grips with it.

Square roots

Square roots: an alternative to saying '3 squared is 9' is to say '3 is the *square root* of 9'. This is written $3 = \sqrt{9}$.

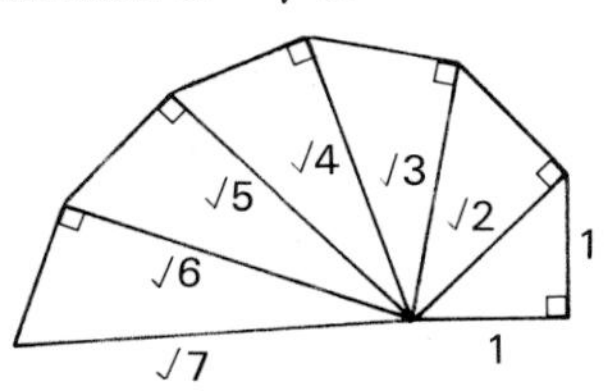

Square roots can be generated as a sort of spiral. Each of the triangles are right-angled and the lengths are calculated by Pythagoras' theorem; for example, in the fifth triangle round, the longest side is $\sqrt{6}$ because $1^2 + (\sqrt{5})^2 = 1 + 5 = 6$.

We have once or twice needed rather awkward phrases like 'a number x such that $x^2 = 2$'. This clumsiness can be avoided by introducing a special term. If x is a positive number such that $x^2 = 2$, we say that x is a *square root* of 2 and we write $x = \sqrt{2}$. (You may have noticed the phrase 'if x is a *positive* number' and wondered why it was inserted. The answer is that it is there simply for our present convenience: there are some special things that we want to say about negative numbers in the next chapter. We shall see that each positive number has a positive square root and a negative square root; for the moment, we can get by with just the positive one.)

It is an interesting and rather elegant fact that the square root of a *whole* number can be only one of two possibilities – it is either itself a whole number or it is irrational. For example the square root of 9 is 3 and the square root of 16 is 4. The square roots of all the whole numbers between 9 and 16 are irrational numbers lying between 3 and 4.

Paper tearing

Have you ever wondered as you walk around a stationery shop with those deliciously smelling piles of papers, rubbers, and pencils just what the paper sizes – A4, A3, A9, and so on mean? Are they purely arbitrary or is there some system in the sequence of sizes?

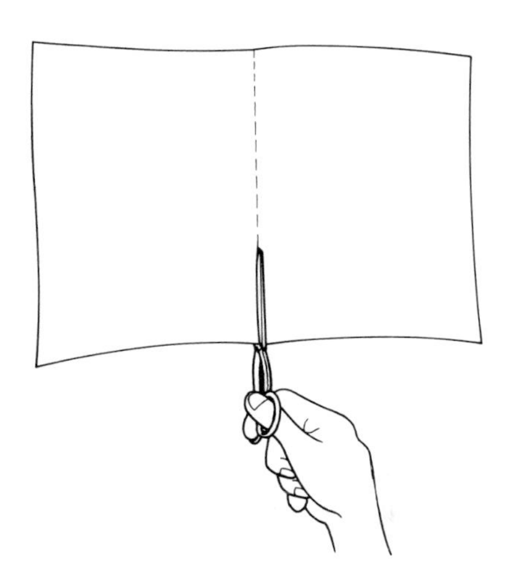

In fact there is a system and it is rather elegant. The largest size sheet of paper in the system is called A0. The next is A1, then A2, and so on. Each size is obtained from its predecessor by cutting it in half. But that's not all there is to it.

1.34

Although the *sizes* get smaller and smaller, the *shapes* stay the same. What does that mean? The different sizes are all exact enlargements or reductions in size, without distortion, just as you would get by enlarging a photograph. If you pick up the A1 piece of paper and turn it round, it is an exact reduction of the A0 size.

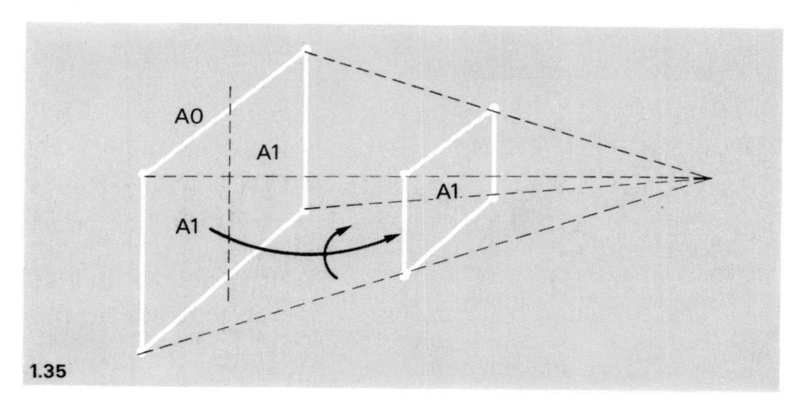

1.35

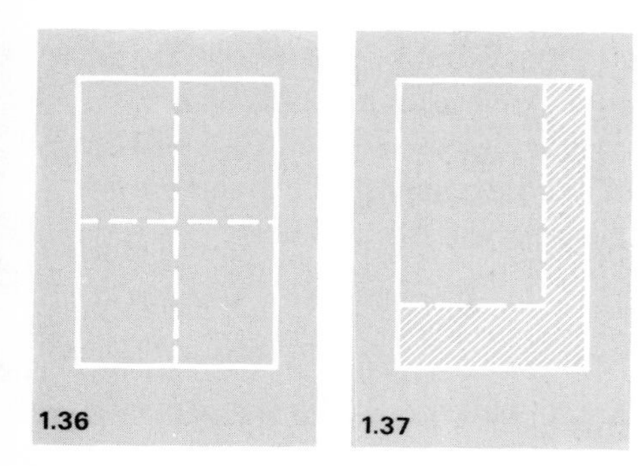
1.36

1.37

You may well say that this is not so hard to achieve. And you would be right. You could just take the next size to be half the length and half the width. But that would give rather a drastic reduction in size – each size would be four times the next one down and there would not be sufficient variety in the sizes. And if you tried to cut a larger piece than this but still of the same shape then there would be a lot of waste.

That's where the idea of just cutting in half comes from. There is obviously no waste. But the shapes of the various sizes might vary quite markedly. For example, if you started with a large sheet of paper measuring 2 metres by 1 metre you would end up with two square pieces, 1 metre by 1 metre, or two long thin pieces 2 metres by $\frac{1}{2}$ metre.

So the problem is to find a shape to start with such that, if it is cut in half, the two halves each have the same shape as the large piece. Unless you have an exceptional mathematical insight you will not be able to guess at the answer. We have introduced this problem here not just because of its interest, but because it is relevant to some of the things we have been saying about the use of symbols – to solve this problem, we *have* to use symbols.

The first difficulty when using symbols is always to decide just what things to represent by symbols. This requires clear thinking about the problem in hand.

What is our problem? We need to find the shape of the paper that we start with. Assuming that the paper is rectangular, its shape is decided by the ratio of the lengths of the sides. So if we start with a sheet of paper measuring x metres by y metres, the ratio $\frac{y}{x}$ determines the shape of the rectangle – long and thin, square, or whatever.

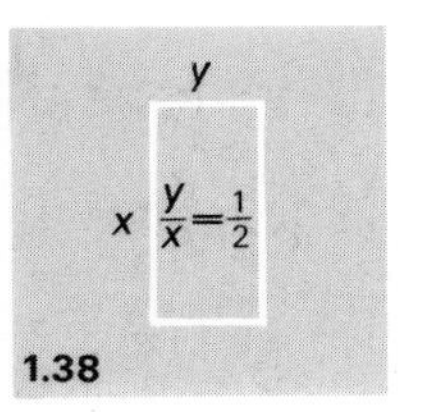

1.38

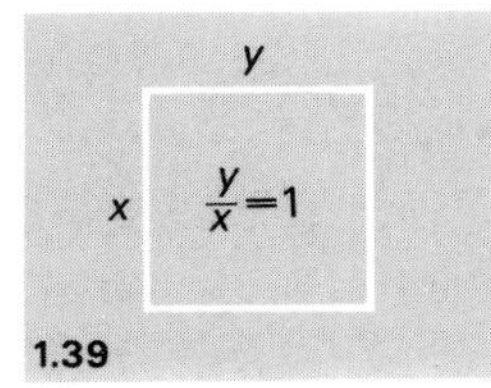

1.39

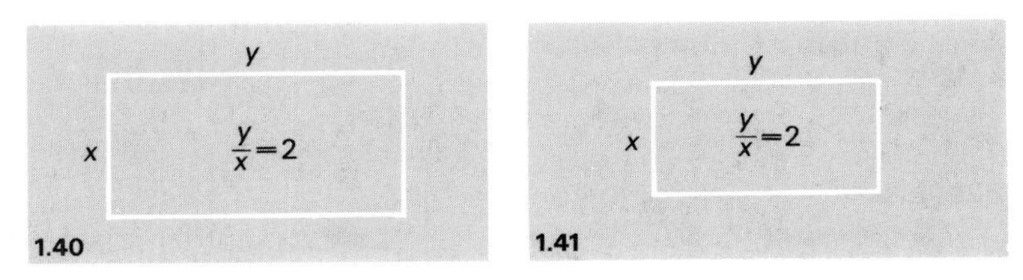

1.40 1.41

If the paper is cut in half, the two pieces will each have sides y metres long and $\frac{x}{2}$ metres long.

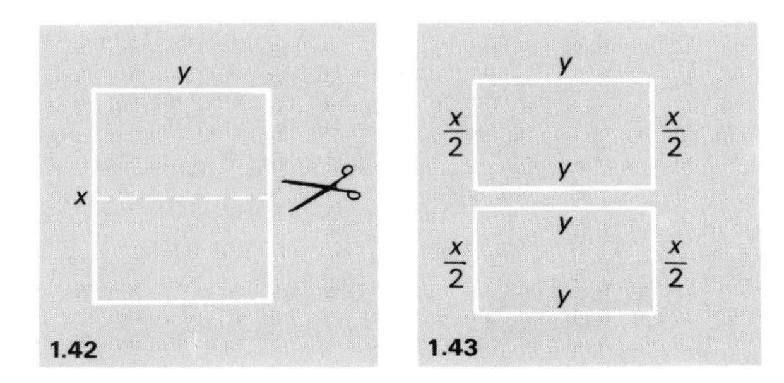

1.42 1.43

(The paper could have been cut the other way, giving two pieces measuring x by $\frac{y}{2}$, it is just a question of settling on one or the other.)

Having chosen what to represent by symbols, we have to translate the problem into the terms of these symbols. The shape of the new pieces will again be determined by the ratio of the lengths of the sides. What is the ratio? It is $\frac{y}{x/2}$ or, if you turn the paper round, it is $\frac{x/2}{y}$.

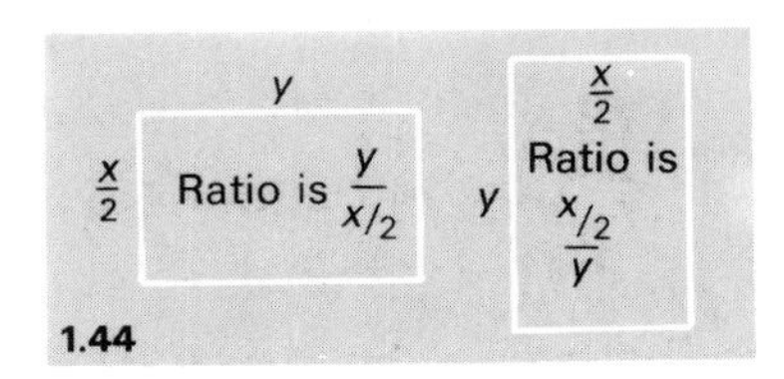

1.44

You may be able to see that you *do* have to turn the paper round. Let us accept that for the moment. If we want the smaller sheet to be the same shape as the larger, we have to choose x and y so that

$$\frac{y}{x} = \frac{x/2}{y}.$$

If we multiply each side of this equation by 2, to get rid of the fraction $x/2$, we get

$$\frac{2y}{x} = \frac{x}{y}.$$

Now remember what we are trying to find: a shape to fit the problem. And the shape is determined by the ratio $\frac{x}{y}$ (or y/x) So let us try to calculate $\frac{x}{y}$. If we multiply each side of the new equation by x we get

$$2y = \frac{x}{y} \times x$$

and if we now divide each side by y we get

$$2 = \frac{x}{y} \times \frac{x}{y}.$$

In other words

$$2 = \left(\frac{x}{y}\right)^2$$

and so

$$\frac{x}{y} = \sqrt{2}$$

and an old friend, $\sqrt{2}$, rears its head again.

So that is how the paper sizes are worked out. If you start with a rectangular sheet of paper whose longer side is $\sqrt{2}$ times the shorter side and cut it in half parallel to the shorter side then the new piece of paper again has its longer side $\sqrt{2}$ times its shorter side. This conclusion can be checked directly

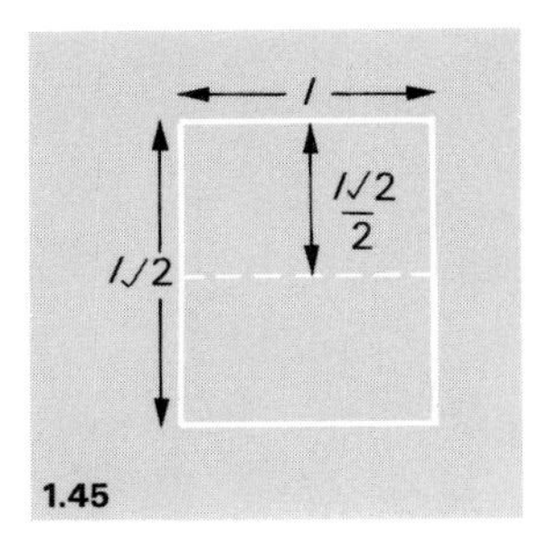

1.45

If the sheet we start with has sides l and $l\sqrt{2}$, the new shorter side is $\frac{l \times \sqrt{2}}{2}$ and so the new longer side is $\left(\frac{l \times \sqrt{2}}{2}\right) \times \sqrt{2}$. We want it to be l, because the new longer side is simply the original shorter side. But

$$\left(\frac{l \times \sqrt{2}}{2}\right) \times \sqrt{2} = \frac{l \times \sqrt{2} \times \sqrt{2}}{2},$$

and $\sqrt{2} \times \sqrt{2}$ is just 2, so the longer side is

$$\frac{l \times 2}{2}$$

which is just l, as advertised!

It may have occurred to you that our answer raises something of a problem. Suppose the paper manufacturers decide to start their paper sizes with a largest piece whose shortest side measures 1 metre. Then the longer side must measure $\sqrt{2}$ metres. But $\sqrt{2}$ is one of those weird irrational numbers. If you cannot express it as the ratio of two whole numbers, then you can never mark off a point '$\sqrt{2}$' on a ruler! We know that it is somewhere between 1 and 2 because 1 is too small ($1^2 = 1$) and 2 is too big ($2^2 = 4$), but just where is anybody's guess. Fortunately, any piece of machinery can operate only to a limited degree of accuracy so we do not need to know $\sqrt{2}$ *exactly*; only close enough. For example, if the cutter operates to within 0.1 mm of accuracy then if we can work out $\sqrt{2}$ to 4 decimal places, we can work out $\sqrt{2}$ metres as accurately as the machine can cope with. Now, here is a subtle point: we shall be seeing later that although you can never calculate $\sqrt{2}$ *exactly*, you can always calculate it as accurately as anybody asks! That is rather a shattering thought but we shall be seeing later in Chapter 5 just what it means.

More paper tearing

When making paper aeroplanes, one often has to start by making a square and the easiest way is to fold over one edge to meet an adjacent edge, cutting off the piece left over.

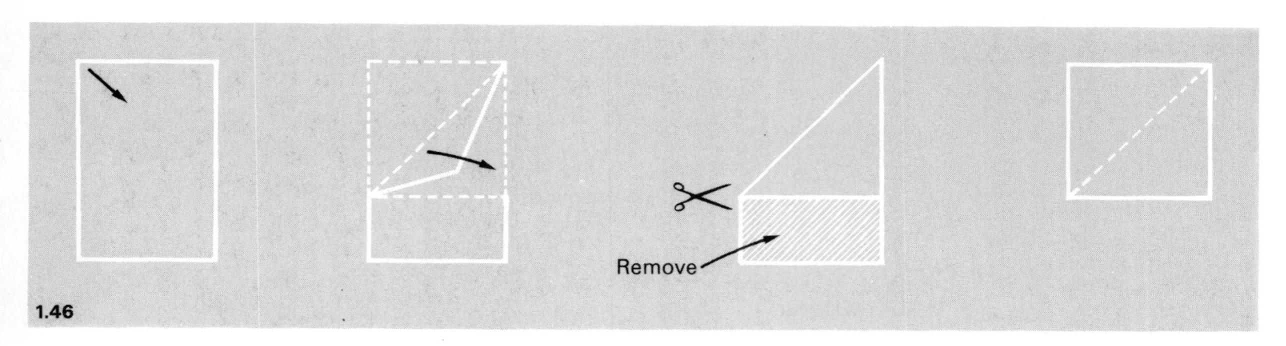

1.46

Now we can pose ourselves a problem very similar to the one of paper sizes. If we started with a piece of paper that is not square then we shall have a piece left over – the piece cut off. It could so happen that this piece has the same shape as the rectangle we started with. The question is what shape do you have to start with so as to guarantee that this will happen?

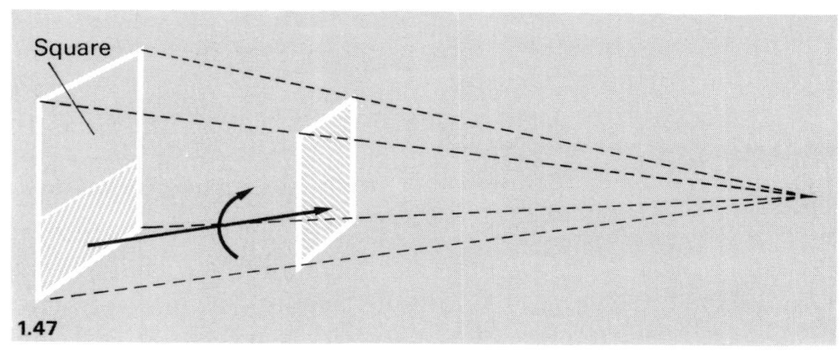

1.47

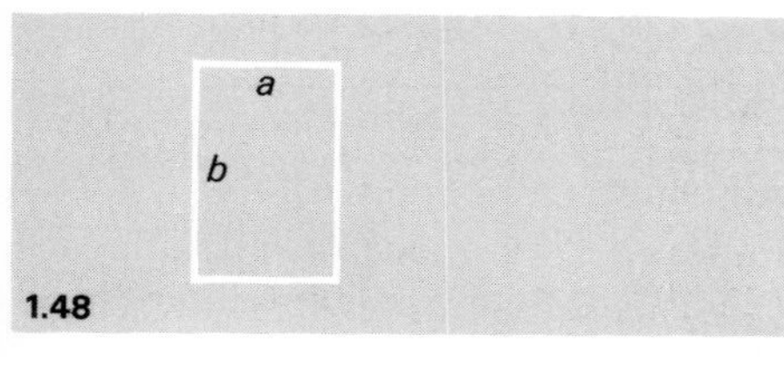

1.48

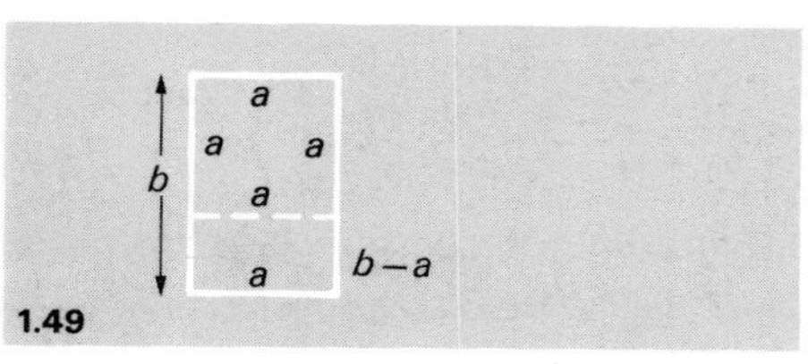

1.49

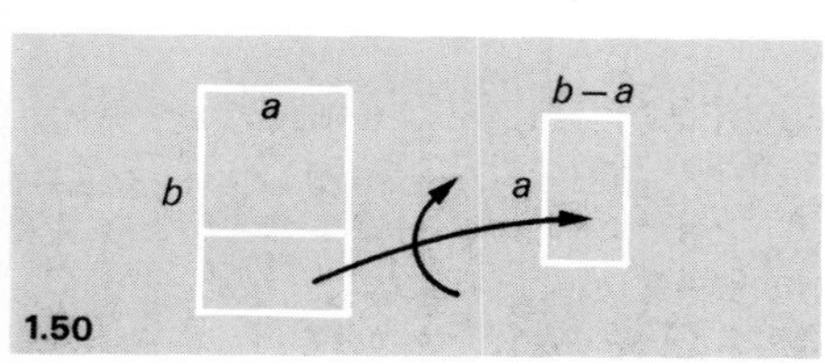

1.50

This problem can be tackled in a similar way to the paper-sizes problem: use symbols to stand for the dimensions of the paper we start with, express the desired property in terms of these symbols and then try to make some deductions.

Suppose the length and width of the paper are a metres and b metres respectively.

Then the dimensions of the smaller piece will be a metres and $b-a$ metres. The shape of the piece we start with is determined by the ratio $\frac{a}{b}$ and that of the remaining piece by $\frac{b-a}{a}$ (again we assume that the piece cut off is to be turned round).

For the shapes to be the same, we must have $\frac{a}{b} = \frac{b-a}{a}$.

We can write $\frac{b-a}{a}$ in a different way. It is $\frac{b}{a}-\frac{a}{a}$, which is just $\frac{b}{a}-1$. (Try it with a few numbers, say $b = 6$ and $a = 2$: work out $\frac{6-2}{2}$ and $\frac{6}{2}-1$). So we want to choose the ratio $\frac{b}{a}$ so that

$$\frac{a}{b}=\frac{b}{a}-1.$$

We can now use the fact that

$$\frac{b}{a}\times\frac{a}{b}=1$$

to get a condition that must be satisfied by $\frac{b}{a}$. If we multiply each side of our equation by $\frac{b}{a}$, we get

$$\left(\frac{a}{b}\right)\times\left(\frac{b}{a}\right)=\left(\frac{b}{a}\right)\times\left(\frac{b}{a}\right)-\left(\frac{b}{a}\right).$$

In other words

$$1=\left(\frac{b}{a}\right)^2-\left(\frac{b}{a}\right).$$

So a and b must be chosen so that $\left(\frac{b}{a}\right)^2$ is the same number as $1+\frac{b}{a}$. This problem is not so daunting as it may seem but it is outside our present purpose, so we shall just quote the answer. It turns out that a and b must be chosen so that the ratio $\frac{b}{a}$ is $\frac{1}{2}+\frac{1}{2}\sqrt{5}$. The presence of $\sqrt{5}$ indicates that again we end with an irrational number. (Remember that we said earlier that the square root of a positive whole number is either a whole number or an irrational number. $\sqrt{5}$ is an irrational number between 2 and 3, because 2×2 is 4 – less than 5 – and 3×3 is 9 – greater than 5.)

This example again illustrates the use of symbols and the occurrence of irrational numbers. But the irrational number we ended up with, $\frac{1}{2}+\frac{1}{2}\sqrt{5}$, though it may look rather unexceptional, is a rather special one. It is called the *golden ratio* and it has a considerable fascination.

If the longer side of a rectangle is $\frac{1}{2}+\frac{1}{2}\sqrt{5}$ times the length of the shorter side, the rectangle left over if a square is cut off is the same shape as the original rectangle. The rectangle can be filled out with squares, the remaining piece at any stage (shaded) being a rectangle of the same shape as the original.

1.51

The golden ratio

The golden ratio appears as a more direct ratio as follows.

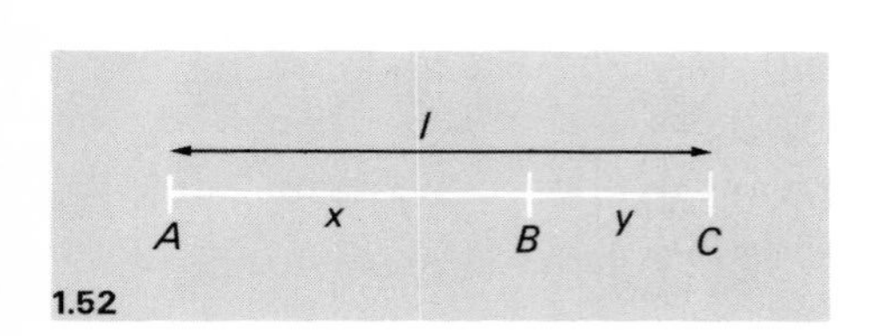

1.52

If a line segment AC is specified, a point B on the line can be thought of as dividing the line into two parts. Suppose the length of AC is l, the length of AB is x and the length of BC is y and that B is positioned so that

$$\frac{l}{x} = \frac{x}{y}.$$

You can see from the diagram that the length l is just $x + y$, so the relationship between l, x, and y takes the form

$$\frac{x+y}{x} = \frac{x}{y}.$$

Replacing $\frac{x+y}{x}$ by the equivalent expression $1 + \frac{y}{x}$ gives

$$1 + \frac{y}{x} = \frac{x}{y}, \text{ which is the same as } \frac{y}{x} = \frac{x}{y} - 1.$$

This is a similar relationship that we got for the paper folding example on page 29: the lengths x and y are in the golden ratio.

The line AC can be bent round the point B to form a rectangle.

1.53

The proportions of the rectangle will depend on the position of B, that is to say on the ratio of the lengths AB and BC.

It is quite remarkable that of all the many occurrences of rectangles in art and architecture, those whose sides are in the golden ratio seem to arise far more frequently than others. Some of these instances were conscious efforts of design.

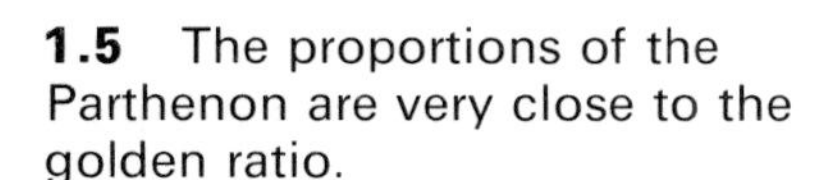

1.5 The proportions of the Parthenon are very close to the golden ratio.

Others were almost certainly subconscious and perhaps these are even more noteworthy: for a certain property to have both an aesthetic and a mathematical appeal, independent of each other, is fascinating to the point of being tantalizing. What is more, it appears that the golden ratio appears quite naturally, independent of human design.

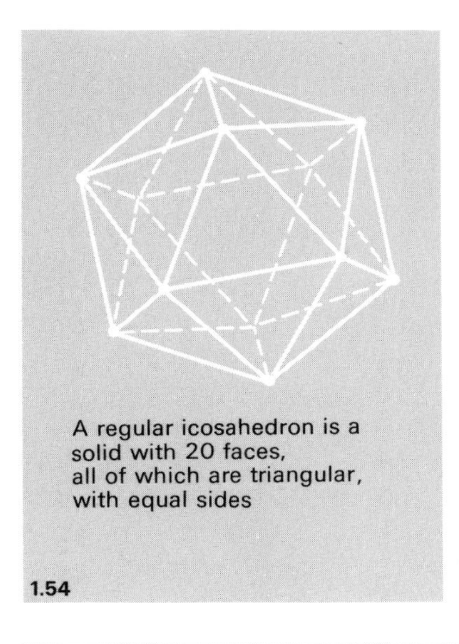

A regular icosahedron is a solid with 20 faces, all of which are triangular, with equal sides

1.54

The vertices can be joined to form rectangles which have sides in the golden ratio!

1.55

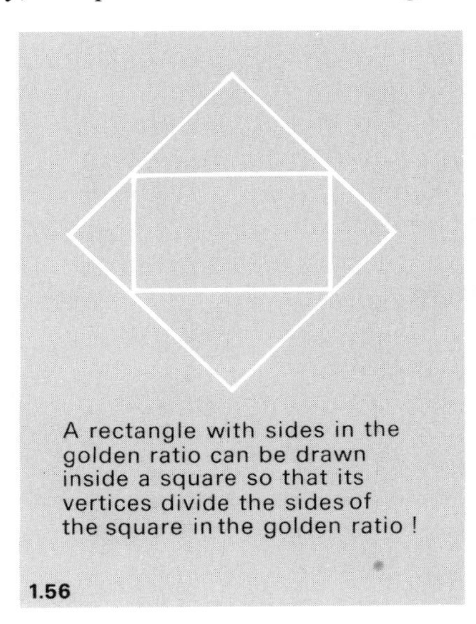

A rectangle with sides in the golden ratio can be drawn inside a square so that its vertices divide the sides of the square in the golden ratio!

1.56

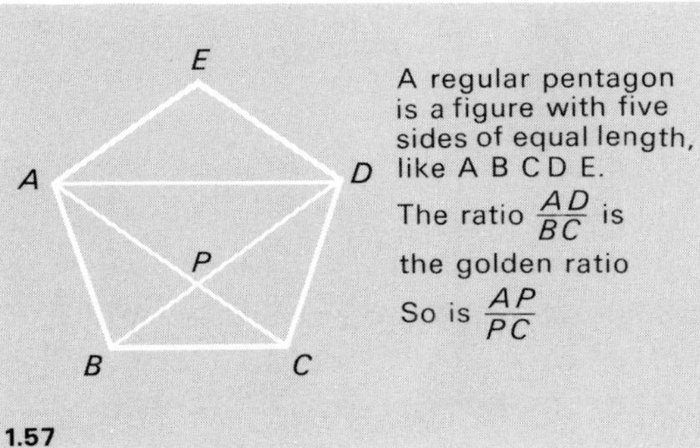

A regular pentagon is a figure with five sides of equal length, like A B C D E. The ratio $\frac{AD}{BC}$ is the golden ratio So is $\frac{AP}{PC}$

1.57

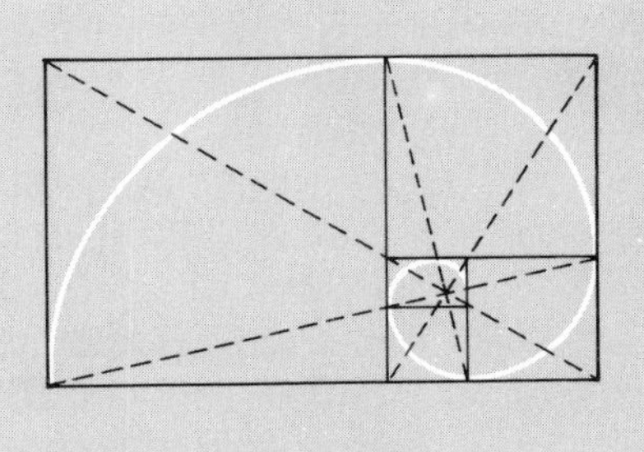

A rectangle with sides in the golden ratio can be separated into squares giving a construction closely connected with spiral common in nature

1.58

Breeding rabbits

Examples of the golden ratio are not confined to geometric contexts. One non-geometric example was investigated in the thirteenth century by Leonardo of Pisa. He posed the problem of calculating the number of rabbits in a breeding colony with the simplification of assuming that a pair of rabbits produces a new pair at the age of two months, and subsequently a new pair each month. We shall assume also that, during the time with which we are concerned, no rabbits die.

Try working out for yourself the number of pairs there would be after the first few months. You should get the following numbers:

1 (at the start), 1 (after 1 month), 2 (after 2 months) 3, 5, 8, 13, 21, ...

Can you spot a pattern in this sequence of numbers?

Each number is the sum of the preceding two.

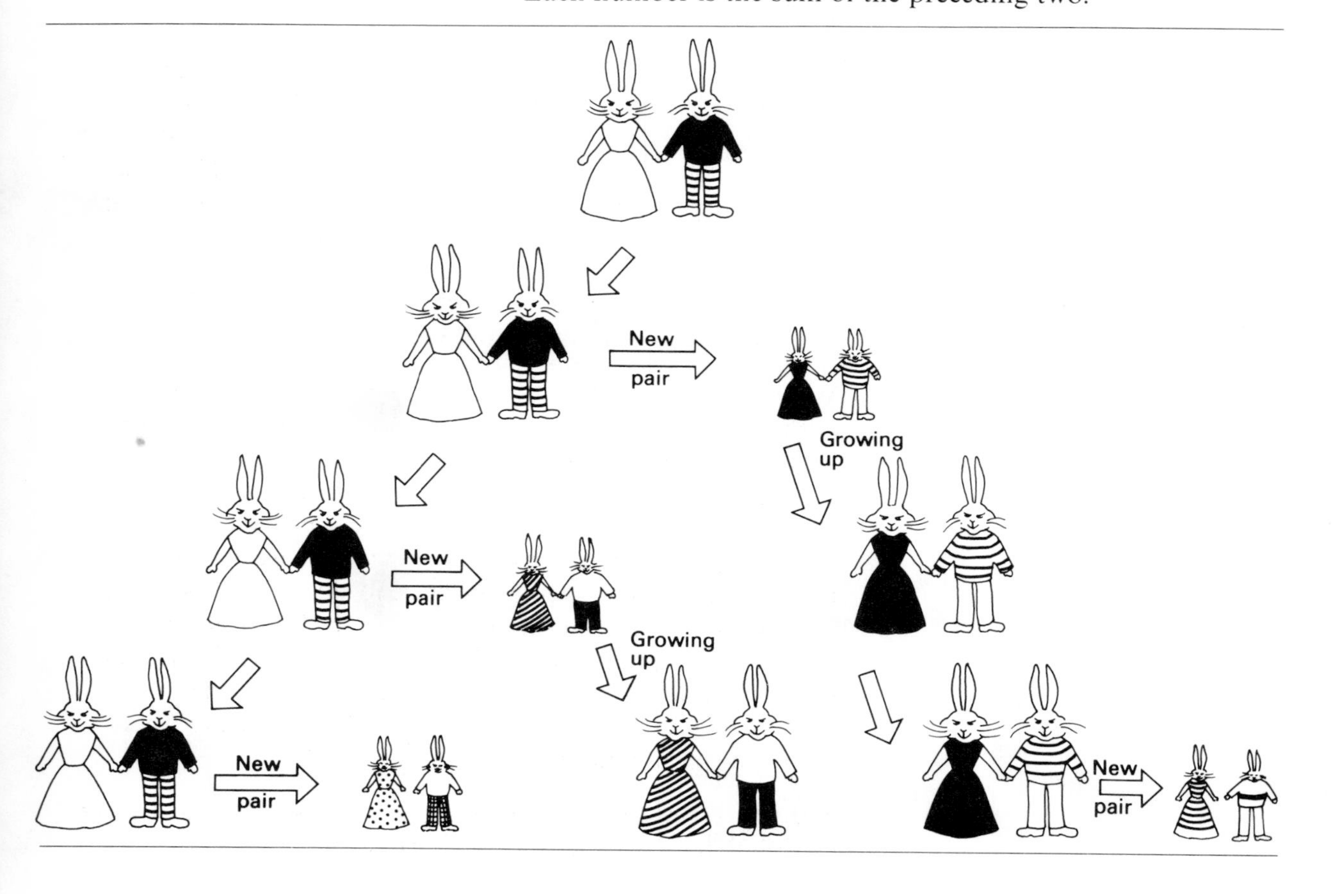

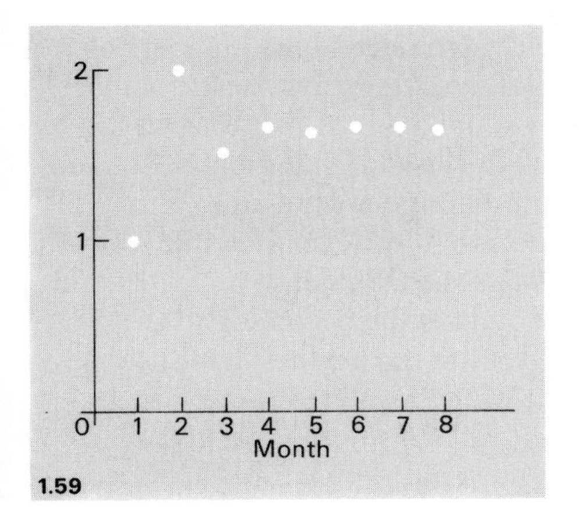

1.59

If you divide each number in this sequence by its predecessor, you get another sequence of numbers. The second number is 1, its predecessor is 1: 1 divided by 1 is 1. The third number is 2; its predecessor is 1; 2 divided by 1 is 2. The fourth number is 3, its predecessor is 2; 3 divided by 2 is 1.5. And so on. The numbers obtained in this way are the ratios of the numbers of pairs of rabbits in successive months and represent a kind of growth rate.

1 1 2 3 5 8 13 21 34 55 . . . .

1 2 1.5 1.667 1.6 1.625 1.6154 1.6190 1.6176 . . . .

The interesting feature of these numbers in the second row is that they cluster together as they progress along the sequence. Because they are clustering closer and closer together, it seems natural to suppose that it may be possible to identify a particular number about which they are clustering – in a sense a sort of *limiting value*. As you might guess, this number is again the *golden ratio*! Can we prove this? Yes, we can, and we begin by introducing some symbols.

As usual, the first problem is to decide what to use our symbols to represent. What is our basic problem about? Not the numbers, 1, 2, 1.5, 1.66 . . ., in the second row, but the numbers of *rabbits* – the numbers in the first row. If we use symbols to stand for the number of rabbits each month, then we can translate our verbal information directly into symbols. Then, when we manipulate the symbols what we will be doing is to manipulate our information – and that is what the use of symbols is all about.

We could use the symbol a to stand for the number of rabbits at the beginning of the first month, b for the second month, c for the third, and so on. But this would have a serious disadvantage. If we want to talk about general properties of the sequence of numbers, we want to be able to refer to a 'general' month. In other words we want to use a symbol to represent a general *position* in the sequence. The symbols a, b, c, . . . do this, but in an unsatisfactory way. Without going through the alphabet on one's fingers, it is not entirely clear what symbol represents the number of rabbits at the beginning of the sixteenth month, say. And what can we do about the fiftieth month?

The way round this is to use a symbol whose numerical value specifies any particular month in question – we shall use n –

Suffix notation: to refer to numbers in a sequence, two symbols are required; one telling us which *position* the number has and one to stand for the number itself. The symbol used for the position, which takes the values 1, 2, 3, ... is usually written as a suffix. Thus we might write r_n to stand for the nth number in a sequence of numbers.

then all we need to do to be able to refer to the numbers of rabbits at the beginning of that month is to associate with n another symbol, r will do, to stand for the thing that we are interested in for the month n – the number of rabbits. So the two symbols together, r and n, will specify the number of rabbits at the beginning of month n. It is usual, in this sort of context, to write the n as a suffix; so we shall write r_n to stand for the number of pairs of rabbits at the beginning of the month n. So the number of pairs at the beginning of the first month is r_1, at the beginning of the second month r_2, and so on. We know the values of r_1, r_2, and so on – and given patience we could work out the value of r_n for any specified value of n, say r_{50}, just by repeated addition. But to make observations of a more general nature, as we want to do, the use of the symbols is essential.

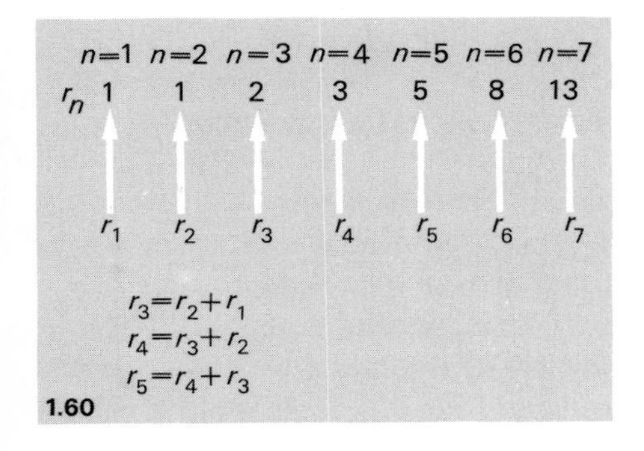

1.60

Having set up the necessary notation, the next step is to translate the problem into the terms of these symbols. So what do we know about r_n? We know that the number of pairs at the beginning of month n is the sum of the numbers of pairs at the beginning of each of the two preceding months, that is to say months $n-1$ and $n-2$. But these numbers are just r_{n-1} and r_{n-2}. Thus the basic rule governing the numbers in the sequence:

Number of pairs = Sum of number of pairs at each of two preceding months

becomes

Number of pairs at beginning of month n = Number at beginning of month $n-1$ + Number at beginning of month $n-2$.

which becomes

$$r_n = r_{n-1} + r_{n-2}.$$

We are now in a position to manipulate the information by performing arithmetic operations on the symbols. You may recall that our original interest was to investigate the ratio of the numbers at month n and month $n-1$, that is to say we wanted to find out something about $\frac{r_n}{r_{n-1}}$. We can do this by dividing each side of our general relation by r_{n-1}: this gives

$$\frac{r_n}{r_{n-1}} = \frac{r_{n-1}}{r_{n-1}} + \frac{r_{n-2}}{r_{n-1}}.$$

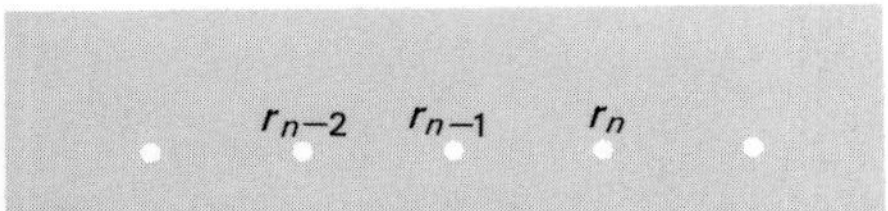

In other words

$$\frac{r_n}{r_{n-1}} = 1 + \frac{r_{n-2}}{r_{n-1}}.$$

r_{n-2} r_{n-1} r_n

1.61

The term on the left-hand side, $\frac{r_n}{r_{n-1}}$, is just what we are interested in, the ratio of the number at month n to the number at the preceding month. The term $\frac{r_{n-2}}{r_{n-1}}$ is not quite the sort of thing that we are interested in; it is the ratio of two successive numbers, but the wrong way round – the number underneath, r_{n-1}, *follows* r_{n-2} rather than precedes it. But all is not lost; we can write $\frac{r_{n-2}}{r_{n-1}}$ in another way. What do we know about two fractions one of which is just the other written upside down, say $\frac{3}{4}$ and $\frac{4}{3}$? If we multiply them together we simply get 1; for example $\frac{3}{4} \times \frac{4}{3}$ is just 1. If we use a letter, K say, to stand for $\frac{3}{4}$, then what is $\frac{4}{3}$ in terms of K? It is just $\frac{1}{K}$, because $K \times \frac{1}{K} = 1$.

We are interested in $\frac{r_{n-1}}{r_{n-2}}$, not $\frac{r_{n-2}}{r_{n-1}}$, and so it makes sense to write

$$K = \frac{r_{n-1}}{r_{n-2}},$$

then, turning the fraction upside down,

$$\frac{r_{n-2}}{r_{n-1}} = \frac{1}{K}.$$

Then our relationship

$$\frac{r_n}{r_{n-1}} = 1 + \frac{r_{n-2}}{r_{n-1}}$$

becomes

$$\frac{r_n}{r_{n-1}} = 1 + \frac{1}{K}.$$

Remember that we were investigating the tendency of the ratios to cluster together as we proceed along the sequence. Clustering around some number or other means that the ratios are getting closer and closer to that number. In other words the successive ratios $\frac{r_n}{r_{n-1}}$ and $\frac{r_{n-1}}{r_{n-2}}$ get closer and closer to each other and closer and closer to some number. We have already seen that if we replace $\frac{r_{n-1}}{r_{n-2}}$ by K, then

$$\frac{r_n}{r_{n-1}} = 1 + \frac{1}{K}.$$

If we replace $\frac{r_n}{r_{n-1}}$ by K as well, then we get

$$K = 1 + \frac{1}{K}.$$

Now multiply each side of this equation by K;

$$K^2 = K + 1.$$

When we first introduced the golden ratio, in our second paper tearing problem, we found, on page 29, that the golden ratio $\left(\frac{b}{a}\right)$ had the property

$$\left(\frac{b}{a}\right)^2 = 1 + \left(\frac{b}{a}\right).$$

This is precisely the property that K must have: so our ratios of numbers of rabbits get closer and closer to the golden ratio. In other words, in the long run the population has a growth rate equal to the golden ratio.

The sequence of numbers 1, 1, 2, 3, 5, 8, ... which we have labelled r_n is a famous sequence, called the *Fibonacci sequence*, Fibonacci being the name by which Leonardo of Pisa, the originator of our rabbit problem, was known.

Calculating irrational numbers

Our long diversion into the golden ratio started as just another example, though a rather fascinating one, of an irrational number. But our work on the Fibonacci sequence throws some light on the problem of the *calculation* of an irrational number.

We mentioned earlier that it is not possible to write down an *exact* decimal expression for an irrational number, as we can for example write 0.75 for $\frac{3}{4}$ or 1.125 for $\frac{9}{8}$. What may be possible, though, is to find a *sequence* of numbers which approach the irrational number more and more closely the further along the sequence one goes.

For example, we have seen that if we take the Fibonacci sequence

1, 1, 2, 3, 5, 8, 13, 21, 34, ...

and calculate the successive ratios

1, 2, 1.5, 1.667, 1.6, 1.625, 1.6154, ...

then these ratios get closer and closer to the golden ratio. A closer inspection shows also that two successive ratios 'sandwich' the golden ratio between them. The alternate terms 1, 1.5, 1.6, 1.6154, get larger and larger. The other alternate terms, 2, 1.667, 1.625, ... get smaller and smaller. The golden ratio is squeezed between these two sequences which act like pincers.

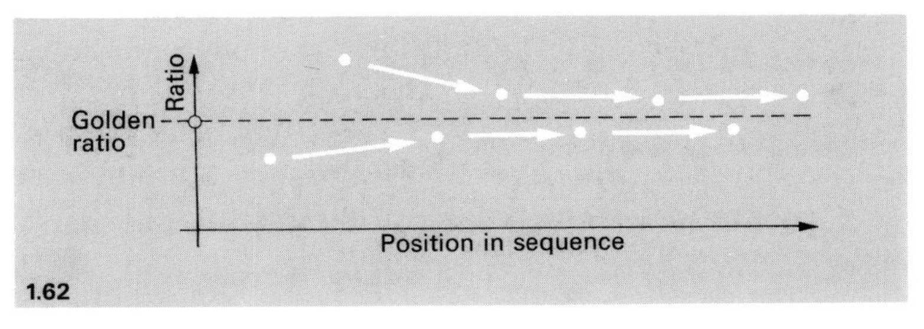

1.62

By going far enough along the terms of the sequences, you can calculate the golden ratio correct to any number of decimal places as desired, though never exactly! For example, we know just from the few terms that we have calculated that the golden ratio lies between 1.625 and 1.6154 and so it has the value 1.6, correct to one decimal place. The next term after 1.6154 is $\frac{34}{21}$ which is 1.6190 ... We can deduce then, that the golden ratio is between 1.615 ... and 1.619 ..., so it is 1.62 correct to two decimal places.

There is one very large question that we have begged in all this. The reasoning all depends on the fact that the ratios $\frac{r_n}{r_{n-1}}$ cluster closer and closer together as n gets larger and larger. To *prove* this is another matter and before that can be attempted, we would have to be more precise as to what we mean by 'cluster closer and closer together'. This notion, of the 'limiting behaviour' of a sequence of numbers is one of the most important and subtle ideas in mathematics. We shall discuss it in Chapter 7.

2 Manipulating symbols

We have seen in the first chapter something of the role played by symbols in mathematics. They are the vital organs of the subject and yet it is often the appearance of masses of symbols that does more than anything to deter the beginner.

If you want to understand mathematics you have to face up to the fact that you cannot do without symbols. But here is the good news: using symbols is not really as bad as it looks – the key to it is practice and confidence. You do not need some magic talent, but you must be willing to take the plunge: nobody learnt to swim by hanging on to the side of the pool. What is more, you will find that occasionally, when you stop and think just what complicated ideas you are manipulating, you will get quite a heady feeling of exhilaration.

Although symbols are used to stand for a great variety of things in mathematics, most of the important ideas can be illustrated using symbols which are used to represent numbers. Consider the difference between the following two statements.

> A rectangular field 100 metres long and 20 metres wide has area 100×20 square metres.
>
> A rectangular field a metres long and b metres wide has area $a \times b$ square metres.

The first statement is essentially arithmetic; it tells you how to calculate a particular area. The second is algebraic; it is a statement about a whole class of situations; it tells you how to calculate the area of *any* rectangular field. Any particular field can be dealt with by replacing a and b by the appropriate numbers.

Rules for manipulating symbols

If we are to use symbols to stand for numbers, then it makes sense to use expressions such as $a \times b$ and $a + b$ and more complicated forms, perhaps involving more symbols. We have already done this to some extent in the previous chapter; what we want to do now is to arrive at a set of rules for manipulating symbols so that we have an agreed record to refer to.

What, for example, does $a+(b+c)-\dfrac{2a\times a\times c-a\times a}{a}$ mean?

There is only one thing to fall back on if we want to answer this. If the symbols stand for numbers, then they must be manipulated according to the arithmetic laws that apply to numbers.

Conventions of arithmetic

When writing an arithmetic expression, conventions have to be agreed upon governing the order in which the expression is to be processed. For example, is $3+2\times 5$ equal to 13 (multiplying first and then adding) or is it 25 (adding and then multiplying)? The convention adopted is that multiplication takes precedence over addition; so $3+2\times 5$ is unambiguously 13.

Brackets are used to indicate a calculation that is to be regarded as an entity: $(3+6)\times 2$ is 9×2 not $3+6\times 2$, which is $3+12$.

If we actually intend the addition to be performed before the multiplication we have to indicate the fact, and that is done by using brackets. Anything enclosed in brackets is to be regarded as a single unit and calculated as a whole. So $(3+2)\times 5$ is 25.

The other two arithmetic operations, subtraction and division, present no further difficulty. Division by 3, for example, can be regarded as multiplication by $\frac{1}{3}$ and subtraction of, say, 5 can be regarded as addition of -5. So division has the same status as multiplication and subtraction has the same status as addition

If additions and subtractions appear together then the convention is to work from the left. For example $3-2+5$ is $1+5$, not $3-7$. Similarly, $3\div 2\times 7$ is $(\frac{3}{2})\times 7$ which is $\frac{21}{2}$; it is not $3\div(2\times 7)$, which is $\frac{3}{14}$. We shall not use the sign $\div$ for division very often, but will usually write a/b or $\dfrac{a}{b}$ for $a\div b$.

The following expressions are unambiguous. You may like to cover up the right-hand half of the page, work out the expressions, and then see if you get the right answer:

$2+(3-2)/4$	$(=2+\frac{1}{4}=2\frac{1}{4})$
$(2-3)/4-2$	$(=-\frac{1}{4}-2=-2\frac{1}{4})$
$2-3/4-2$	$(=-\frac{3}{4})$
$(2-3)/(4-2)$	$(=-1/2)$
$2+3\times 2\div 4$	$(=2+6\div 4=2+6/4=3\frac{1}{2})$

Properties of arithmetic operations

There are other properties of the arithmetic operations that are used when performing calculations. They are rather different from the conventions that we have seen so far in that they are intrinsic to the operations themselves, rather than a matter of avoiding ambiguity in writing. These properties may well appear unimportant to you but, if so, that is only because they are so familiar to you. In fact, they are important enough to be given names, for in more abstract situations, when symbols are used to stand for things other than numbers, it is the absence or presence of properties such as these that helps with the comparison between the new system and the familiar structure of the numbers. You will be seeing something more of this in Chapter 4.

Addition

The order in which the numbers are written is immaterial, for example $1+2$ and $2+1$ are the same. In general,

$$a+b=b+a.$$

The property is referred to by saying that addition is *commutative.*

If three numbers have to be added together, it does not matter how they are paired. For example $(1+3)+5$ is the same as $1+(3+5)$. In general,

$$(a+b)+c=a+(b+c),$$

the brackets are redundant, and we can write $a+b+c$ without ambiguity.

This property is referred to by saying that addition is *associative.*

Multiplication

Multiplication is also

commutative, $a \times b = b \times a$

and

associative, $(a \times b) \times c = a \times (b \times c)$

(so we can write $a \times b \times c$ without ambiguity).

As we said before, these may seem trivial and obvious, but in a way that is precisely the point: we are trying to move from a familiar concrete situation to an abstract environment. The first step in this process is to translate well-known and comfortable properties into the new language.

When using symbols, it is usual to drop the multiplication sign and write ab instead of $a \times b$. This is a convenient shorthand which is possible with symbols, but not with numbers – there would be confusion between, for example, 23 (meaning twenty-three) and 2×3. The commutative and associative laws now look like this:

$$ab = ba, \quad (ab)c = a(bc) = abc.$$

Interaction of multiplication and addition

The laws of arithmetic

$$\left.\begin{array}{l} a+b=b+a \\ ab=ba \end{array}\right\}$$

$$\left.\begin{array}{l} a+(b+c)=(a+b)+c \\ a(bc)=(ab)c \\ \text{so we can just write} \\ a+b+c \text{ and } abc \end{array}\right\}$$

$$a(b+c)=ab+ac$$

An expression such as $2 \times (3+4)$ can be calculated in two ways. The obvious way is to calculate 2×7 directly, to give 14. But it is also possible to perform multiplications first: $2 \times 3 + 2 \times 4$ also gives the correct answer, 14. In general, we have

$$a(b+c) = ab + ac$$

and we say that *multiplication is distributive over addition.*

It is important to realize that there is nothing deep about this: there is nothing extra really to understand except that this is a sensible way to manipulate symbols if they are to model ordinary arithmetic. If you need any convincing about the distributive property, it may help to see it in the context of calculating areas. A rectangle with sides a metres and $(b+c)$ metres has area $a(b+c)$.

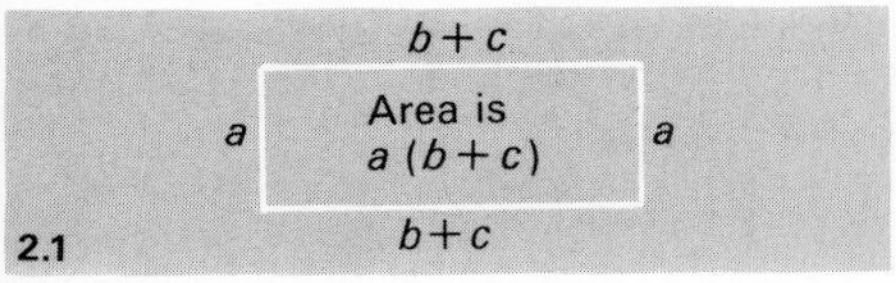

$b+c$

a Area is $a(b+c)$ a

$b+c$

2.1

b c

a Area is ab Area is ac

b c

Total area is $ab+ac$

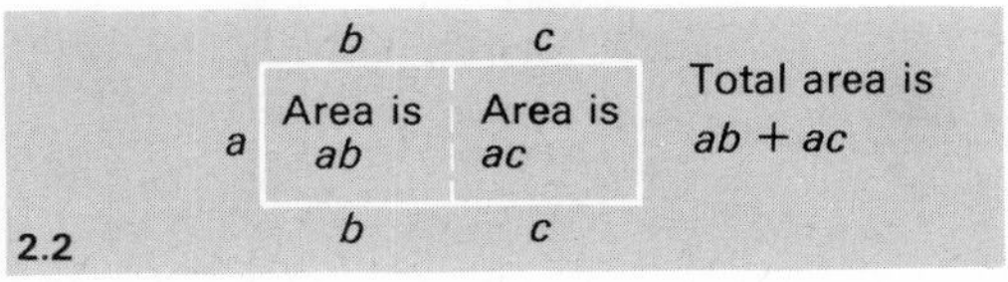

2.2

This is obviously the same as the sum of the areas of two rectangles of sides a and b metres and a and c metres respectively.

Notice that addition is not distributive over multiplication: $2+(3\times 6)$ is not the same as $(2+3)\times(2+6)$.

This completes the properties of addition and multiplication that we need for the manipulation of symbols. It is not at all necessary to remember the names given to the properties and, although the list does appear rather long and uninspiring, you will soon realize that they are just the rules that you have already used subconciously in ordinary calculations.

A useful example of the use of the property that multiplication is distributive over addition is its application to an expression such as $(x+y)^2$. It is often useful to express this in an alternative form. Remember that

$$(x+y)^2 \text{ means } (x+y)(x+y).$$

How can we apply the distributive law? It tells us something about expressions like $a(b+c)$ and that is not quite what we have here. Why not let the left hand $(x+y)$ play the role of our a and then the other x and y be our b and c? Like this:

$$\begin{array}{ccccccc} (x+y) & (x+y) & = & (x+y)x & + & (x+y)y \\ \uparrow & \uparrow\ \uparrow & & \uparrow\ \ \uparrow & & \uparrow\ \ \uparrow \\ a & (b+c) & = & a\ \ b & + & a\ \ c \end{array}.$$

Now, because multiplication is commutative, $(x+y)x = x(x+y)$ and $(x+y)y = y(x+y)$, so the right-hand side that we've just worked out is the same as

$$x(x+y)+y(x+y).$$

Now we can use the distributive property again, on each of these terms. This gives us

$$x^2+xy+yx+y^2.$$

Since we know that xy is the same as yx, then this is the same as

$$x^2+2xy+y^2.$$

So

$$(x+y)^2 = x^2+2xy+y^2.$$

You might like to work out $(x+y)^3$ for yourself – regard it as $(x+y)(x+y)^2$: you should get

$$(x+y)^3 = x^3+3x^2y+3y^2x+y^3.$$

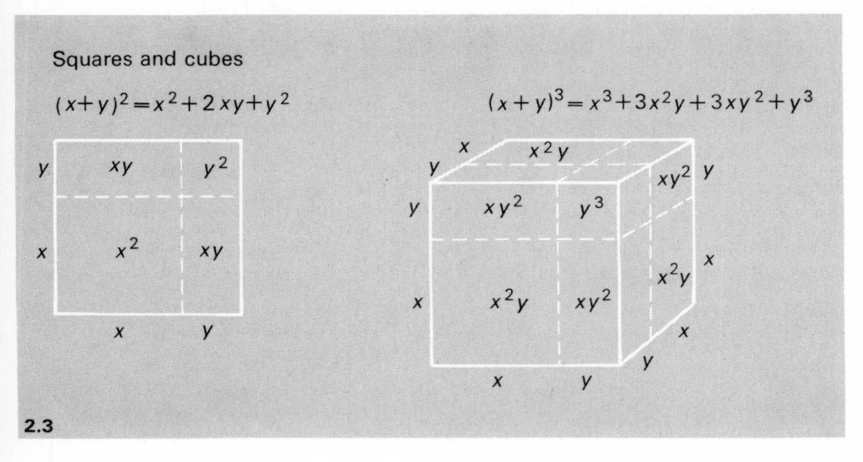

2.3

It remains only to complete this catalogue by listing the properties of subtraction and division. The order in which two numbers are subtracted or divided *does* make a difference to the answer – for example, $2-4$ is not the same as $4-2$ and $2 \div 3$ is not the same as $3 \div 2$. So neither subtraction nor division is commutative. So

$$a-b \text{ is not the same as } b-a.$$

But if these expressions are not equal, just how are they related? Quite simply, actually; it turns out that

$$a-b = -(b-a).$$

(For example, $6-3 = 3$ and $3-6 = -3$.)

And for division,

$$\frac{a}{b} \text{ is not the same as } \frac{b}{a};$$

in fact $\frac{b}{a} = \frac{1}{a/b}$.

(For example, $\frac{2}{3} = \frac{1}{3/2}$.)

We also have problems with subtraction and division when we try to combine three numbers. The expression $a-b-c$ is ambiguous, because we get a different answer if we associate the a and the b together first to that which we get if we associate the b and the c together first.

Minus signs
Remember that, for example
$3-(-2)$ is $3+2$
$3+(-2)$ is $3-2$
$(-3)\times(-2)$ is 6
$3\times(-2)$ is -6.
See p.

For example $8-5-2$ is ambiguous. Is it $(8-5)-2$, which is $3-2$, just 1? Or is it $8-(5-2)$, which is $8-3$, that is to say 5? Again we ask, if

$$(a-b)-c \text{ is not equal to } a-(b-c)$$

is there any other way of writing them? Well, we don't need the brackets in $(a-b)-c$ because we have the convention of working through an expression from left to right: the convention tells us to interpret

$$(a-b)-c \text{ as } a-b-c\,.$$

One way of understanding that

$-1\times-1=+1$

is to think of numbers as lying on a line and of multiplying by -1 as 'reflecting' in 0.

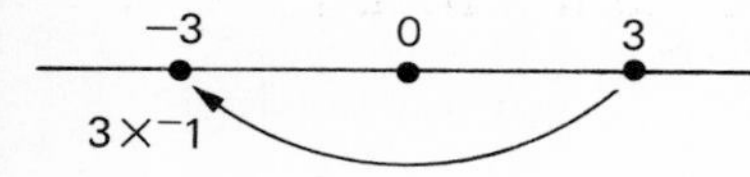

So a number to the left of 0 (negative) gets reflected to one to the right (positive).

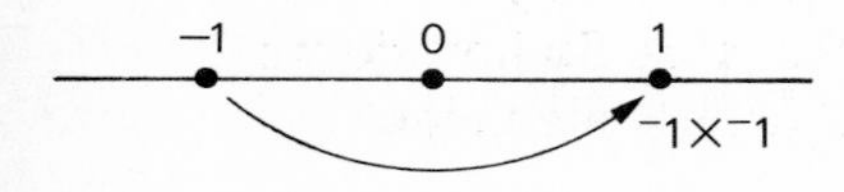

But can we reinterpret $a-(b-c)$? Can we remove the brackets? Well, it turns out that

$$a-(b-c)=a-b+c\,.$$

Try it for yourself – convince yourself it's true by trying a few numbers for a, b, and c. For example, if $a=6$, $b=3$, and $c=1$,

$$\begin{aligned}a-(b-c)&=6-(3-1)\\&=6-2\\&=4\end{aligned}$$

and

$$\begin{aligned}a-b+c&=6-3+1\\&=3+1\\&=4\,.\end{aligned}$$

Similar problems arise when you try to combine three numbers by division, because

$$(a/b)/c \text{ is not the same as } a/(b/c)\,.$$

(For example $(12/3)/2$ is $4/2$, which is 2, but $12/(3/2)$ is $12/1.5$ which is 8.) It is possible to remove the brackets though:

$$(a/b)/c=a/bc$$

and $a/(b/c)=ac/b$.

Frequently, one is interested in the magnitude of a number, irrespective of its sign. For example, -3 and $+3$ have the same magnitude. We write $|a|$ for the magnitude of a number a. So $|-3|=3$, $|5|=5$, and so on, $|a|$ is called 'the modulus of a'.

You will probably find it time well spent to go through these statements on the last few pages with a, b, and c replaced by particular numbers, say $a=2$, $b=3$, $c=4$. You may also feel it worthwhile in the light of our list of properties to look back over any of the calculations in Chapter 1 that you may have been worried about on first reading.

To give you some practice in manipulating symbols, we shall now look at a few problems that involve juggling with symbols, but which are interesting in their own right and which we shall be finding useful later on.

Adding squares

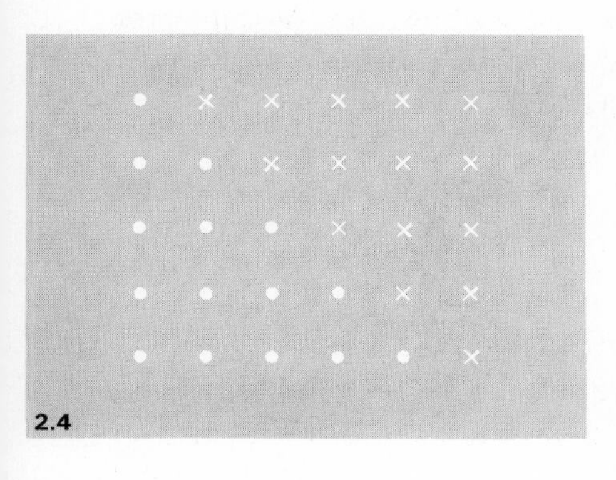
2.4

The first problem is similar to those we tackled in the previous chapter concerning sums of numbers. Do you remember the result,

$$1+2+3+\ \dots\ +n=\frac{n(n+1)}{2}$$

and similar results for adding odd and even numbers? We pictured these results by representing the numbers as dots and the sums as arrays of dots. Building up a succession of lines of dots gives a two-dimensional array of dots – and that is why each of the results contains terms involving n^2, reminiscent of calculating areas.

Can we apply the same idea to

$$1^2+2^2+3^2+\ \dots\ +n^2,$$

the sum of the *squares* of the first n whole numbers?

This time, *each number* can be represented by an *array* of dots, and thought of in terms of an area. So how do we represent the sum? By a stack of areas – a volume.

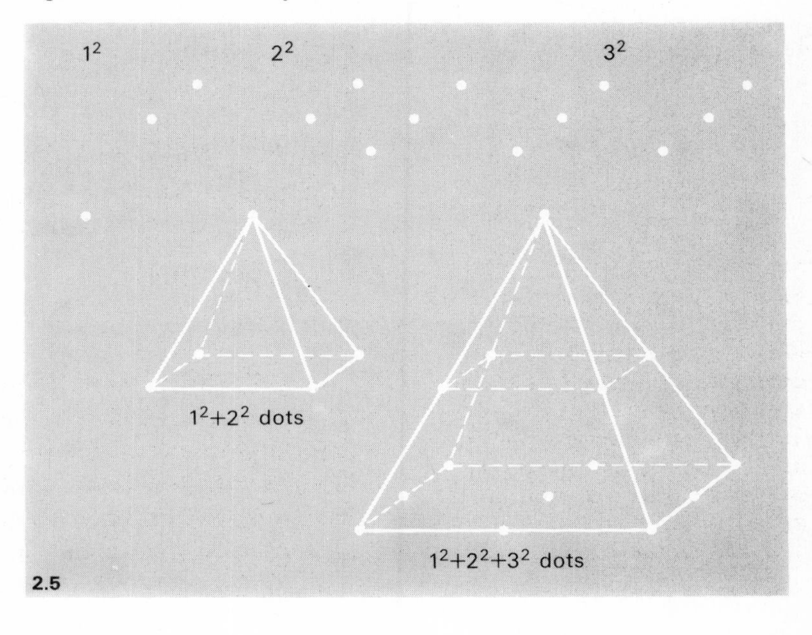

2.5

Each sum; 1^2, 1^2+2^2, $1^2+2^2+3^2$, and so on, can be represented by a pyramid of dots. For example, the sum $1^2+2^2+3^2$ has a 3×3 base with the next layer 2×2, topped by the 1×1 vertex. The sum $1^1+2^2+3^2+\ldots+n^2$ has n layers, starting with an $n \times n$ base.

When we worked out the sum $1+2+3+\ldots+n$ on p. 18, we were able to calculate it from the area of a square (although it was not exactly half the area – every dot down the diagonal had to be counted, they were not shared out equally between the two halves). It is not easy with the pyramid; it is some fraction of the volume of a cube, but what fraction? Here is a cunning way to work it out.

$(x+y)^3$ means $(x+y)(x+y)(x+y)$ – the number 3 tells you how many times $(x+y)$ appears, in the multiplication. Similarly, n^3 means $n \times n \times n$, n^4 means $n \times n \times n \times n$, and so on.

Since the number we are looking for is some fraction of the volume of a cube, it must involve n^3 in some way, let's say there is a term an^3 in the answer – in much the same way that there is a $\frac{1}{2}n^2$ in the earlier formula. There may also be some n^2 terms and some terms in n. Let's propose, then, that the sum of the first n squares has the form

$$an^3+bn^2+cn+d,$$

where a, b, c, and d, are numbers that we have to find.

We can dispose of d straight away. When n is zero, this expression has the value d, but for n equal to zero we are considering the sum of the first zero squares, which is just zero! So d is zero, and our proposition becomes

$$1^2+2^2+3^2+\ldots+n^2 = an^3+bn^2+cn$$

The way in which we disposed of d is a clue as to the way we can calculate a, b, and c – we gave n a special value. If the formula is to hold true for all values of n, then by replacing n by any particular value we get a condition that must be satisfied by a, b, and c. For example, setting n equal to 1 gives

$$1^2 = 1^3a+1^2b+1c$$

in other words

$$a+b+c = 1.$$

The value of a, b, and c that we seek must satisfy this condition.

With $n = 2$, we get

$$1^2+2^2 = 2^3a+2^2b+2c$$

in other words

$$8a+4b+2c = 5.$$

For $n = 3$, we get

$$27a+9b+3c = 14.$$

We now have three conditions to be satisfied by the three numbers a, b, and c that we seek and these three conditions are sufficient to specify those numbers precisely. The method by which conditions such as these are manipulated to produce the required values is an interesting topic in its own right, but to go into details now would take us too far along a sidetrack. We shall resist being seduced and simply quote the result: it turns out that

$$a = \tfrac{1}{3},\ b = \tfrac{1}{2},\ \text{and } c = \tfrac{1}{6};$$

you can check them if you want to by substituting these three values into each of the conditions.

So we now have the result we sought:

$$1^2+2^2+3^2+\ \dots\ +n^2 = \tfrac{1}{3}n^3+\tfrac{1}{2}n^2+\tfrac{1}{6}n.$$

Do you think that this demonstration constitutes a proof? Don't worry about the way that the values of a, b, and c were pulled out of the bag: they certainly satisfy the conditions. There is a more serious objection. We have assumed that the answer has a particular form, an^3+bn^2+cn, and have *proved* that *if* it has this form *then* it must be $\tfrac{1}{3}n^3+\tfrac{1}{2}n^2+\tfrac{1}{6}n$. Of course, we did not just pull this form out of thin air, we produced some fairly plausible reasoning and it is arguable that this does in fact constitute a proof. We *know* that our answer is true for n equal to 0, 1, 2, and 3 and we could *check* it for *any* value of n. But we could not check it for *every* value of n. Our method would not convince a real sceptic that the result is true for *all* positive whole numbers.

But having got a result to work with, it is possible to produce more convincing methods of proof. This is a very common method of working in mathematics: one produces a heuristic argument to obtain a result in which one has great confidence and *then* sets out to prove it. There is little future in attempting a proof from the outset without knowing what result you want to prove. In any case there are more methods available for proving the validity of a result than to combine a proof that a result is true at the same time as deriving that result.

Here is one way to prove particular results which concern whole numbers, as does our present one.

An interesting method of proof

The method is essentially the method of contradiction that we used in the previous chapter – but it is interesting in the way it can be especially adapted to proving results of this type, concerning the whole numbers.

The proposition to prove is that, whatever positive whole number we assign to n,

$$1^2+2^2+\ldots+n^2=\tfrac{1}{3}n^3+\tfrac{1}{2}n^2+\tfrac{1}{6}n.$$

Suppose, on the contrary, that the result were not true. This would mean that there is at least one number for which it is not true. We already know that it *is* true for *some* values of n ($n=1$ for example), so if it is not true for *all* values of n then there must be two adjacent values of n, say $n=k$ and $n=k+1$ such that the result is true for $n=k$ and not true for $n=k+1$. When you have got that straight in your mind, consider the following piece of manipulation.

If the result is true for $n=k$,

$$1^2+2^2+\ldots+k^2=\tfrac{1}{3}k^3+\tfrac{1}{2}k^2+\tfrac{1}{6}k.$$

Now add $(k+1)^2$ to each side

$$1^2+2^2+\ldots+k^2+(k+1)^2=\tfrac{1}{3}k^3+\tfrac{1}{2}k^2+\tfrac{1}{6}k+(k+1)^2$$

But

$$(k+1)^2=k^2+2k+1.$$

$(x+y)^2=x^2+2xy+y^2$
so
$(k+1)^2=k^2+2k+1$

(We worked out $(x+y)^2$ earlier, on page 42; this result follows by putting $x=k$ and $y=1$.)

So the right-hand side is

$$\tfrac{1}{3}k^3+\tfrac{1}{2}k^2+\tfrac{1}{6}k+k^2+2k+1.$$

Now let's rearrange this expression. It is the same as

$$\tfrac{1}{3}k^3+k^2+k+\tfrac{1}{3}+\tfrac{1}{2}k^2+k+\tfrac{1}{2}+\tfrac{1}{6}k+\tfrac{1}{6}.$$

This may seem arbitrary – you will see why we have done it shortly and you will also see that it was by no means a flash of insight to do it – there was a perfectly good reason. But check the terms – the ones in k^3, the ones in k^2, those in k, and the ones not involving k – and you will see that the two expressions are the same.

Now the group of terms

$$\tfrac{1}{2}k^2+k+\tfrac{1}{2}$$

can be expressed differently. They amount to

$$\tfrac{1}{2}(k^2+2k+1)$$

which is

$$\tfrac{1}{2}(k+1)^2 .$$

In much the same way,

$$\tfrac{1}{3}k^3+k^2+k+\tfrac{1}{3}=\tfrac{1}{3}(k^3+3k^2+3k+1)$$
$$=\tfrac{1}{3}(k+1)^3 .$$

$(x+y)^3 = x^3+3x^2y+3xy^2+y^3$
so
$(k = 1)^3 = k^3+3k^2+3k+1$

(We saw an expression for $(x+y)^3$ earlier, on page 43.) So our total expression, which is $\tfrac{1}{3}k^3+k^2+k+\tfrac{1}{3}+\tfrac{1}{2}k^2+k+\tfrac{1}{2}+\tfrac{1}{6}k+\tfrac{1}{6}$, can be rewritten as

$$\tfrac{1}{3}(k+1)^3+\tfrac{1}{2}(k+1)^2+\tfrac{1}{6}.$$

What have we proved then? That if

$$1^2+2^2+3^2+\ \dots\ +k^2=\tfrac{1}{3}k^3+\tfrac{1}{2}k^2+\tfrac{1}{6},\text{ then}$$

$$1^2+2^2+3^2+\ \dots\ +k^2+(k+1)^2=$$
$$\tfrac{1}{3}(k+1)^3+\tfrac{1}{2}(k+1)^2+\tfrac{1}{6}.$$

This is our formula with n replaced by $k+1$. So we have *proved* that if the formula is true with n replaced by k it *must* be true with n replaced by $k+1$.

True for any n?
1. Suppose it is not true.
2. There must be a number k such that true for k but not for $k+1$.
3. *Prove* that if true for k *must* be true for $k+1$.
4. Contradiction. So assumption in 1 is false.
5. True for every n.

In other words, there *cannot* be two consecutive values of n, $n=k$ and $n=k+1$, such that the result is true for $n=k$ and not true for $n=k+1$. This contradicts our original assumption that the result is not true for every positive whole number. Our conclusion, then, is that the result *is* true for every number:

$$1^2+2^2+3^2+\ \dots\ +n^2=\tfrac{1}{3}n^3+\tfrac{1}{2}n^2+\tfrac{1}{6}n .$$

The volume of a pyramid

It was some way back that we started the problem of the sum of squares, but perhaps you will recall that we found it useful to picture the sum as a collection of dots arranged in the form of a pyramid. We are now going to set out actually to calculate the volume of a pyramid and the method we shall use will introduce an important idea – the notion of an *infinite process*. Such processes are vital to the understanding of the calculus and the concept of a limit – one of the most subtle and exciting ideas in mathematics and a topic that we shall be investigating in Chapter 7.

The method is hardly new: it was used in the fourth century BC by the Greek philosopher Democritus. But as with so much of the work done by the Greeks, it touches on some very deep ideas. And that is why we are taking up this particular one, because it introduces some ideas which we shall be developing in later chapters. The idea is to get a sequence of approximations to the answer and use this sequence to predict what the answer must be!

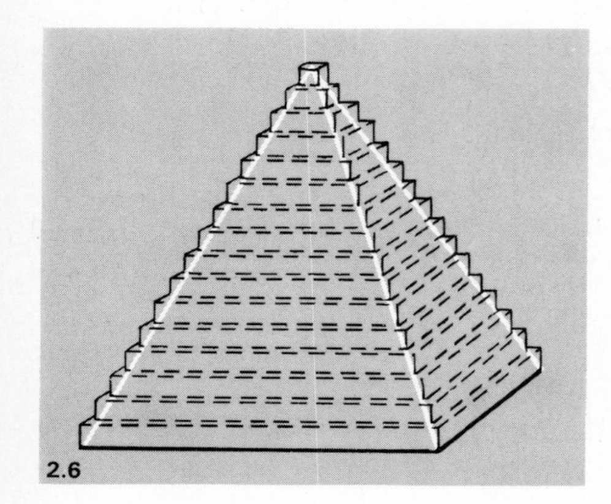
2.6

This is how the approximations are calculated: by imagining a stack of square plates tapering in the same way that the pyramid does.

It is easy to work out the volume of a square plate if you know its dimensions, so it seems feasible to work out the total volume of the stack of plates. This would give an approximate value for the volume of the pyramid.

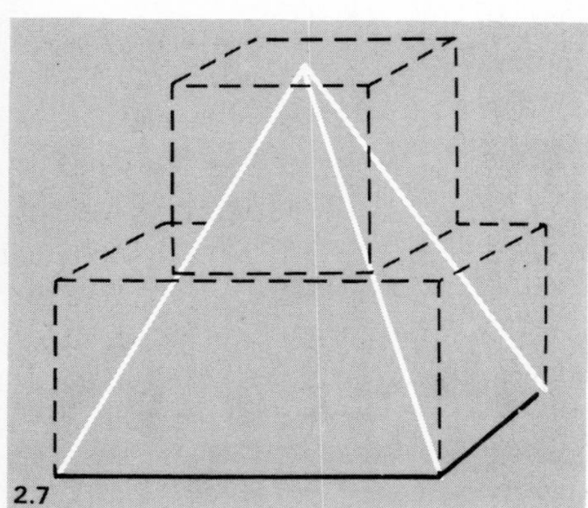
2.7

If we used only two plates, the approximation would be very crude. With four, we would get an improvement. With four thousand, we would feel that we must be very close to the answer. By successively reducing the thicknesses we get a sequence of approximations which get closer and closer to the volume we seek.

If the dimensions of the pyramid were known, we could calculate each of the approximations and, when the change in two successive approximations is small enough to be neglected, then we can take the latest approximation as a good estimate of the volume of the pyramid. For example, if we wanted to perform the calculation correct to two places of decimals, then as soon as successive approximations differred by less than 0.005, we could feel confident that no further improvement in the approximation would affect the first two decimal places.

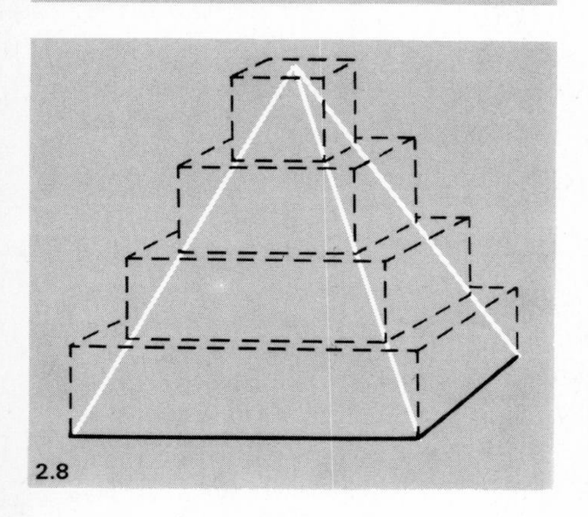
2.8

Such a numerical calculation can be applied to any pyramid whose dimensions are known; it is essentially an arithmetic process that has to be carried out afresh for each new case. If we could use symbols to specify a general pyramid, and if we could duplicate our approximation procedure in terms of these symbols, then we may be able to obtain a *formula* which could be applied to *any* pyramid directly.

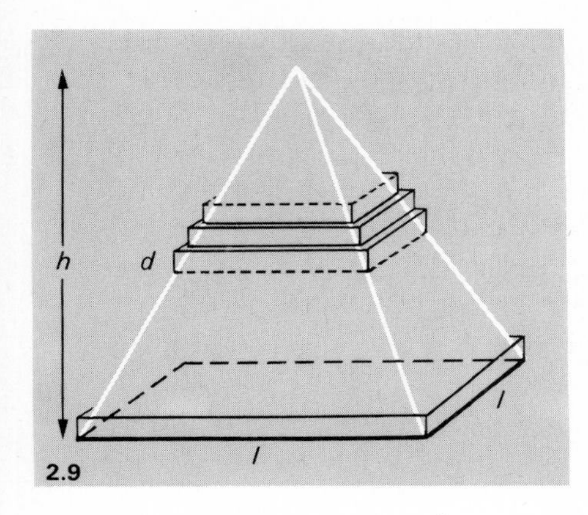

2.9

There are a number of ways of specifying a pyramid. We shall assume that we are given its height and the length of the sides of its base. So let us consider a pyramid with height h and whose square base has sides of length l, and let the thickness of each of the square slabs be d. Having specified the problem, we can get down to some detailed work. Be warned, though; we are about to embark on some quite sophisticated symbol manipulation. You will not find it easy, but try to follow the general drift of the argument. And remember, this is not a 'second-best' approach – this is exactly what mathematicians do themselves when first meeting a new and complicated piece of work.

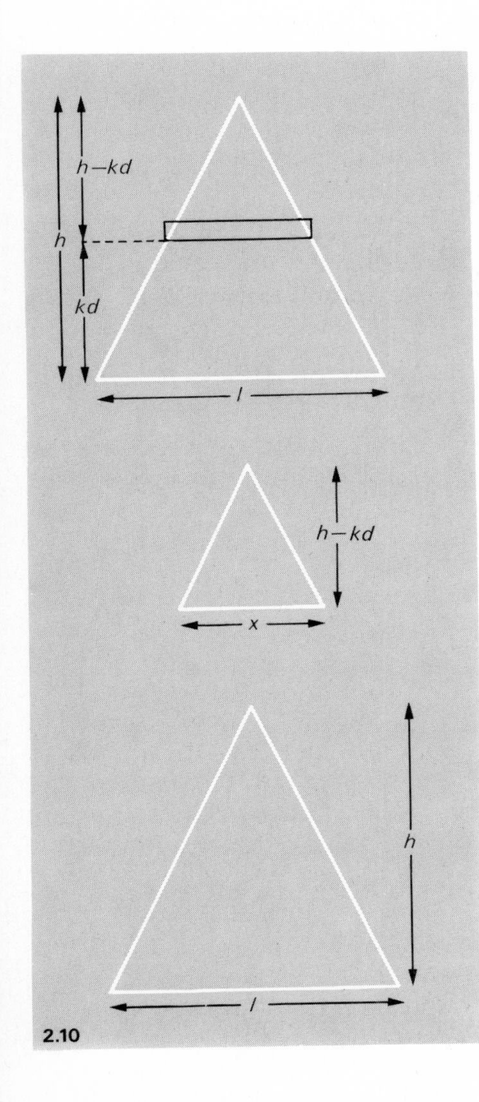

2.10

To work out the dimensions of slabs higher up the stack we need first of all to introduce a symbol that we can use to refer to an arbitrary slab – we can use the symbol k and talk about the kth slab from the bottom. How do we work out the size of this slab?

If it is the kth slab from the bottom, then its distance from the base of the pyramid must be kd.

You can think of it as the base of a smaller pyramid (a piece of the original) with height $h-kd$. Now the large and the small pyramids are the same shape – they are in the same proportions – and so the ratio of the length of base to height must be the same for each pyramid. For the large pyramid the ratio is l/h. What is it for the small one? The length of base of the small pyramid is just what we are trying to find – so let's call it x. Its height is $h-kd$. So the ratio in question is $x/(h-kd)$. The fact that the two pyramids have the same proportion is expressed symbolically as

$$\frac{x}{h-kd}=\frac{l}{h}.$$

(Do you remember the problem on paper sizes in Chapter 1?)

In other words

$$x=\frac{l(h-kd)}{h}.$$

The volume of this kth slab is x^2d, so it is

$$\frac{l^2(h-kd)^2d}{h^2}.$$

To work out the total volume of the slabs, we have to add together the volumes of the first ($k = 1$), second ($k = 2$) and so on. Let's say there are n slabs: then the volume is

1st slab	2nd slab	3rd slab		nth slab
$\frac{l^2 d}{h^2}(h-d)^2 +$	$\frac{l^2 d}{h^2}(h-2d)^2 +$	$\frac{l^2 d}{h^2}(h-3d)^2$	$+ \dots +$	$\frac{l^2 d}{h^2}(h-nd)^2$
$(k = 1)$	$(k = 2)$	$(k = 3)$		$(k = n)$

This can be written as

$$\frac{l^2 d}{h^2}[(h-d)^2 + (h-2d)^2 + \dots + (h-nd)^2].$$

(Compare this with $ab + ac = a(b+c)$, where a is $\frac{l^2 d}{h^2}$.)

Now let's work out each of the terms in the large brackets. We know from p. 42 that

$$(x+y)^2 = x^2 + 2xy + y^2$$

$$\text{so } (h-d)^2 = h^2 - 2hd + d^2$$

$$\text{and } (h-2d)^2 = h^2 - 4hd + (2d)^2 \text{ and so on.}$$

Our total volume is therefore

$$\frac{l^2 d}{h^2}[[h^2 - 2hd + d^2] + [h^2 - 4hd + (2d)^2] + [h^2 - 6hd + (3d)^2] + \dots + [h^2 - 2nhd + (nd)^2]].$$

This looks pretty awful, but don't be put off. Every group of three terms contains an h^2; and how many such groups are there? n of them. So we can write nh^2 to take account of all the h^2 terms. If we also do some rearranging of terms, we get

$$\frac{l^2 d}{h^2}[nh^2 - (2hd + 4hd + 6hd + \dots 2nhd) + (d^2 + (2d)^2 + (3d)^2 + \dots + (nd)^2].$$

One of the brackets contains a $2hd$ in every term. The other contains a d^2 in every term. So we can rewrite the volume as

$$\frac{l^2 d}{h^2}[nh^2 - 2hd(1 + 2 + 3 + \dots + n) + d^2(1^2 + 2^2 + 3^2 + \dots + n^2)]$$

Now we are getting somewhere, because we have a formula for $1+2+\ldots+n$, worked out in the previous chapter, and for $1^2+2^2+\ldots+n^2$, worked out earlier in this chapter.

$$1+2+\ldots+n = \tfrac{1}{2}n^2+\tfrac{1}{2}n$$

$$1^2+2^2+3^2+\ldots+n^2 = \tfrac{1}{3}n^3+\tfrac{1}{2}n^2+\tfrac{1}{6}n$$

Using these formulae, we get

$$\frac{l^2d}{h^2}\left[nh^2-hdn^2-hdn+\tfrac{1}{3}d^2n^3+\tfrac{1}{2}d^2n^2+\tfrac{1}{6}d^2n\right].$$

There is a further simplification that we can make. The height of the pyramid is h, the thickness of the slabs is d, and the number of the slabs is n. There must be a relationship between n, h, and d. What is it? n slabs of thickness d, stacked on top of each other, will have a total height of nd. So

$$nd = h.$$

So we can replace h in the formula by nd, or since $d = h/n$, we can replace d by h/n; or we could replace n by h/d! It would not really make sense to replace h, because h is something given to us – the height of the pyramid. But n and d are of our own introduction, so let's replace one of those. It turns out to be simpler to replace n. To do so requires quite a lot of pushing around of symbols – but, once again, don't be deterred: take your time and follow through the working steadily.

If we replace n by h/d whenever it appears we get the following form for the right-hand side of our formula:

$$\frac{l^2d}{h^2}\left[\frac{h^3}{d}-\frac{hdh^2}{d^2}-\frac{hdh}{d}+\frac{1}{3}\frac{d^2h^3}{d^3}+\frac{1}{2}\frac{d^2h^2}{d^2}+\frac{1}{6}\frac{d^2h}{d}\right]$$

which is

$$\frac{l^2d}{h^2}\left[\frac{h^3}{d}-\frac{h^3}{d}-h^2+\frac{1}{3}\frac{h^3}{d}+\frac{1}{2}h^2+\frac{1}{6}hd\right].$$

We can remove the brackets by multiplying each term inside the brackets by $\dfrac{l^2d}{h^2}$. This gives

$$l^2h-l^2h-l^2d+\tfrac{1}{3}l^2h+\tfrac{1}{2}l^2d+\tfrac{1}{6}\frac{l^2d^2}{h} \text{ which is just}$$

$$\tfrac{1}{3}l^2h-\tfrac{1}{2}l^2d+\tfrac{1}{6}\frac{l^2}{h}d^2.$$

Expressing a numerical process algebraically is not just a question of aesthetics–it would be an essential first step if one wished to instruct a computer to perform a calculation.

That may have been a lot of work, but think what it has done for us. If we are given the dimensions of a pyramid – told the values of h and l – we can immediately calculate an approximate value for the volume corresponding to any thickness of slab – any value of d. Our algebraic manipulation, tiresome as it was, has been done once and for all and gives us a result that we can apply directly to *any* pyramid on a square base.

Suppose, for example, we had the case of a pyramid for which l was 4 and h was 7. Then

$$\tfrac{1}{3}l^2h - \tfrac{1}{2}l^2d + \tfrac{1}{6}\frac{l^2}{h}d^2$$

would be

$$\tfrac{1}{3} \times 4^2 \times 7 - \tfrac{1}{2} \times 4^2 \times d + \tfrac{1}{6} \times \frac{4^2}{7} \times d^2,$$

which turns out to be

$$37.33 - 8d + 0.381d^2$$

So with $d = 0.1$, our approximation to the volume is 36.53 and with $d = 0.01$ it is 37.25.

We could use the approximating formula in this way for any values of l and h – any square pyramid – and for any sequence of values of d to give a sequence of approximations to the volume.

By introducing symbols and manipulating them in a formal way, without having to refer back to an interpretation of their meaning we have in fact manipulated information about pyramids in general, and the approximation procedure, which would have been incredibly complex if we were to have attempted the same task using ordinary language. But there is still more that the symbols can do for us.

Remember the role played by d, the thickness of the slabs. The idea is to reduce d to obtain improvements to the approximations. For a given pyramid, h and l are fixed. What happens to $\tfrac{1}{2}ld^2$ and $\tfrac{1}{6}\frac{l^2}{h}d^2$ as d gets smaller? They also become smaller; and the smaller they become, the more closely does our expression approach the value $\tfrac{1}{3}l^2h$. What is more, we can get as close as we please to $\tfrac{1}{3}l^2h$ by choosing d small enough.

What does this amount to then? Surely, $\frac{1}{3}l^2h$ must be a *precise* formula for the volume of the pyramid!

So we have got even more than we hoped for – not just an algebraic description of the approximation process, but an extremely simple formula for the *exact* calculation of the volume.

As we mentioned earlier, this method is essentially the same as that used by Democritus over 2000 years ago and, being a Greek, he was interested not just in the practical result, but the method by which it was obtained; and the method worried him. The surface of the approximating stack is not smooth like the surface of the pyramid and an unevenness persists however thin the slabs. The actual point of transition from an uneven approximation to the smooth pyramid appeared to Democritus to be quite inexplicable. He was right to worry – it was just such a problem as this which was at the heart of the difficulties that Newton and Leibnitz had with the calculus in the seventeenth century. In Chapter 7 we shall see how the mathematicians of the nineteenth century resolved these difficulties.

3 Describing curves

We have seen in the first two chapters something of the way that symbols can be used in mathematical arguments. In this chapter we shall meet one of the most inventive and powerful uses of symbols. We shall see how they can be used to create a link between geometry and algebra; not just a passive, descriptive role, but one which establishes a genuine unifying step, enabling each subject to enlighten and motivate the development of the other.

Identifying points with numbers

The key idea is deceptively simple and certainly not news to anyone who has read a map. The common method of locating a point on a map is to quote the latitude and longitude of the point or, alternatively, a pair of grid references special to the map. This is the essential idea of *coordinate geometry*.

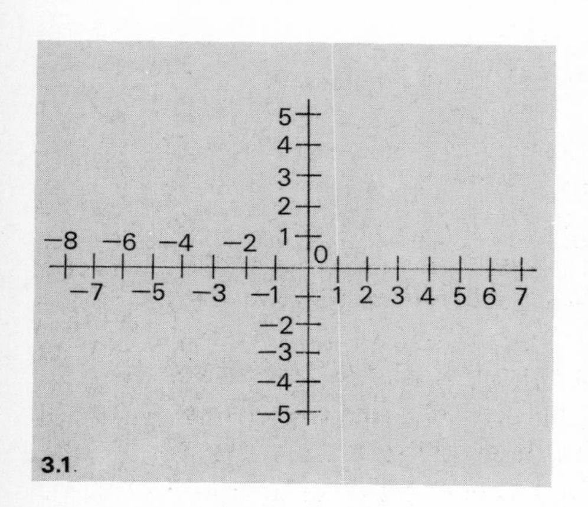

3.1.

If a pair of perpendicular lines are drawn, intersecting at a point 0, they can be used to locate any point in the plane containing the lines. Each point on the lines can be used to represent a number, the distance of the point from 0, with negative numbers being used for points on one side of 0 and positive numbers for points on the other side. The distances between the points does not matter. It could be a millimetre or a mile. It is important only that some scale is agreed, no matter which, just as different scales can be used for maps without prejudicing the idea of a grid system. (This correspondence between numbers and points on a line is more subtle than it may appear, as we saw in Chapter 1: we shall spend some time in Chapter 5 sorting it out but the difficulties do not intrude on the matters we want to discuss in this chapter.)

Any point in the same plane as these two lines can be identified by quoting two distances: the distances you have to move parallel to each of the lines to get to the point from 0. In Fig. 3.2, A is identified by the numbers $2\frac{1}{2}$ and 2, B by 3

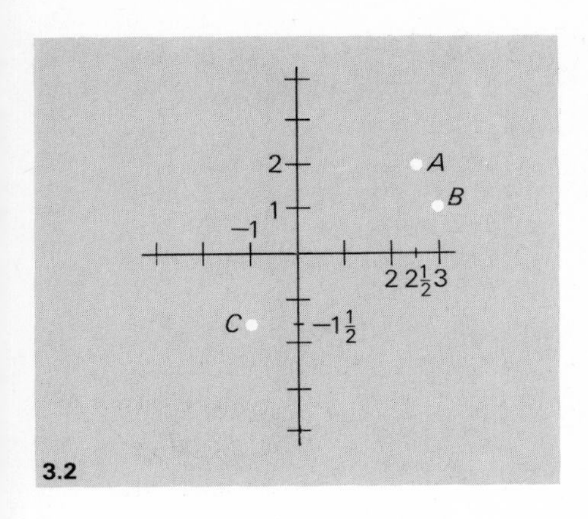

3.2

and 1, and C by -1 and $-1\frac{1}{2}$: notice how negative numbers are used to indicate moves to the left or downwards. Any general point in the plane can be identified by a pair of numbers (x, y) where x is the length of the path parallel to the line running across the page and y is the length of the path parallel to the line running down the page. So the x-value is recorded on the line running across the page and the y-value on the line running down (or up) the page. The choice of the symbols x and y in this way is such a strong convention that the lines are called the x-axis and the y-axis respectively. The particular values taken by x and y for any particular point are called the x- and y-coordinates of that point: in the example in Fig. 3.2, the point A has x-coordinate $2\frac{1}{2}$ and y-coordinate 2.

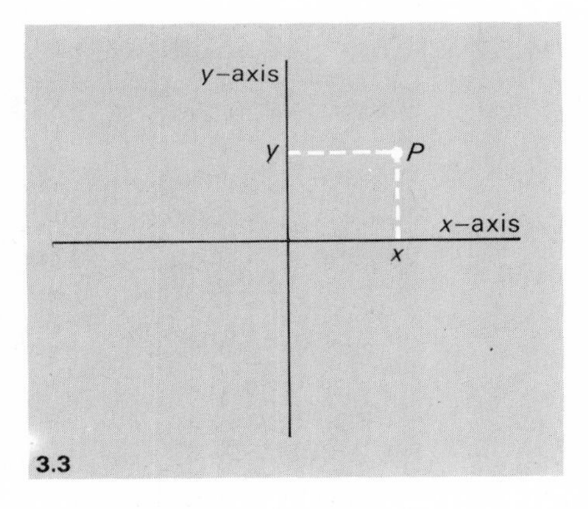

3.3

The procedure may seem somewhat tame, but it is really a significant step: it establishes a correspondence between the set of all points in a plane and the set of all pairs of numbers. Any pair (x, y) specifies a point and any point specifies a pair of numbers. This presents us with the possibility of transferring problems from one system to another. A geometric problem, concerned with points, lines, and so on can be transferred to an algebraic problem, concerned with the manipulation of symbols. On the other hand, an algebraic problem may be capable of a geometric interpretation. In later chapters we shall see more of this idea of the transference of problems from one situation to another – not for nothing do people complain that the first thing a mathematician does when you ask him a problem is to answer a different one!

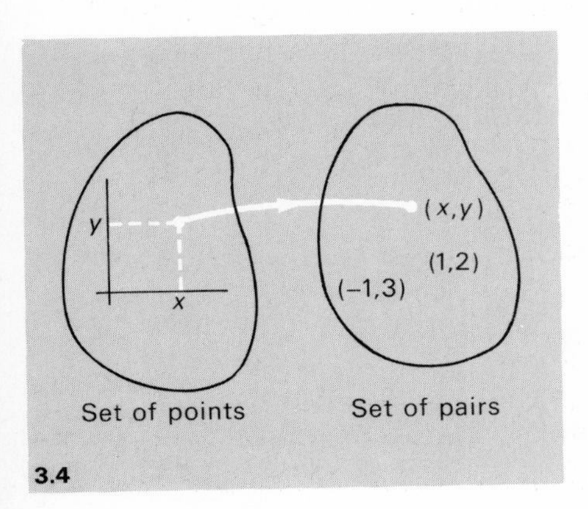

3.4

We have not got the space here to go into the details of the techniques of solving geometric problems using algebra but

René Descartes (1596–1650) developed the notion of using algebraic methods to tackle geometry–the method is called *Cartesian* geometry. He was fortunate in his education. The rector of his Jesuit College allowed him to lie in bed in the morning–a habit Descartes regarded as 'above all conducive to intellectual profit and comfort'! His philosophical works earned him a scandalous reputation amongst the Protestant clergy.

we can see how the descriptive process works – how geometric shapes can be described algebraically. Apart from its intrinsic interest, we are going to need to use the idea later. We shall start with one of fundamental notions of geometry – distance.

Describing distance algebraically

First of all, let us find an algebraic expression for the distance between two points which have the same y-coordinate. Suppose A has coordinates (x_1, y_1) and B has coordinates (x_2, y_1). If this sounds rather abstract and you feel you're going to get lost, look at what this means visually by referring to Fig. 3.5. We are actually dealing with a very simple situation. Remember that x_1 is the distance of A from the y-axis and that x_2 is the distance of B from the y-axis. The line AB is parallel to the x-axis so what is the distance between A and B? Look at Fig. 3.5—can you see that it is just $x_2 - x_1$?

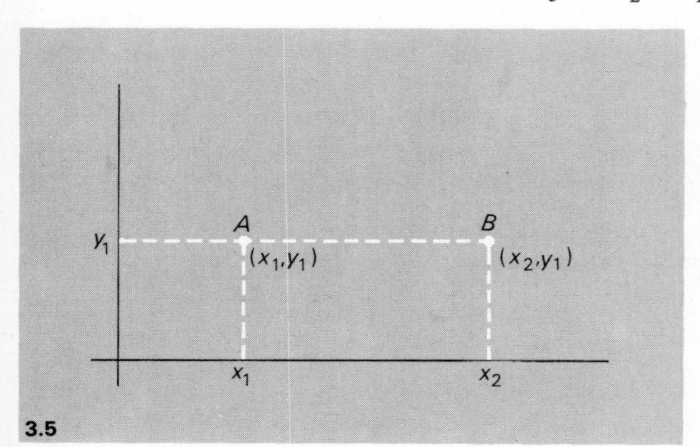

3.5

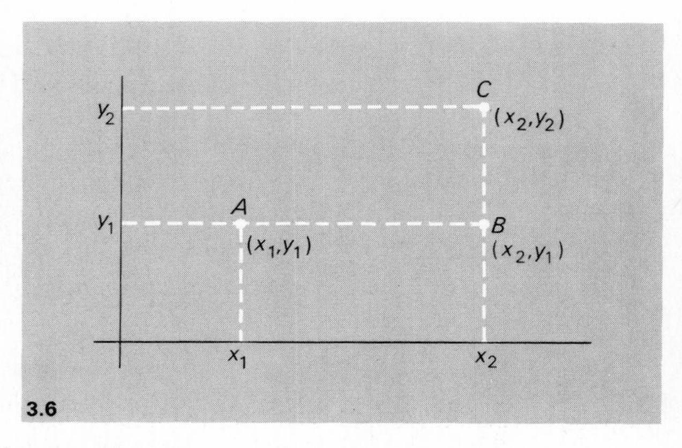

3.6

The same idea applies to two points with the same x-coordinate. If C has coordinates (x_2, y_2), then the distance between B and C is just $y_2 - y_1$.

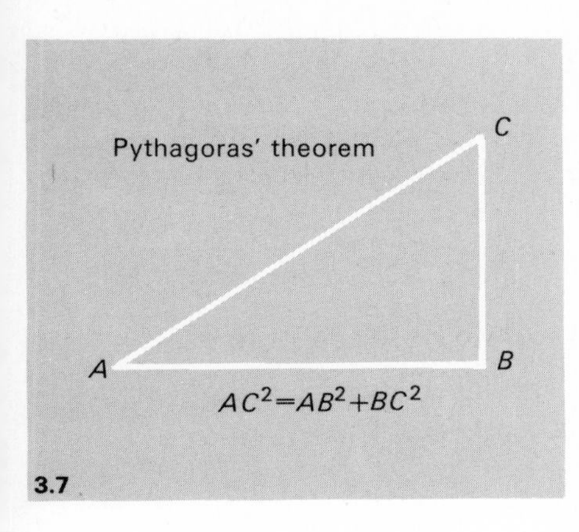

3.7

How can we work out the rather more difficult distance AC? The triangle ABC has a right-angle at B, so we can use Pythagoras' theorem – we met it in Chapter 1. Writing AC for the distance between A and C, and using BC and AB in a similar way, the theorem tells us that

$$AC^2 = AB^2 + BC^2.$$

We have already worked out AB and BC – so we now have

$$AC^2 = (x_2 - x_1)^2 + (y_2 - y_1)^2.$$

Now look at the expression on the right-hand side, remembering that the coordinates of A are (x_1, y_1) and of C (x_2, y_2). The expression on the right involves only the coordinates of A and C: B does not come into it; so the formula tells us how to write down the distance between any two points directly once we know their coordinates. For example, if A has coordinates $(1, 2)$ and C has coordinates $(3, 7)$, then the distance between A and C comes out as follows:

Distance between points.
If A is (x_1, y_1) and C is (x_2, y_2),
$AC^2 = (x_2 - x_1)^2 + (y_2 - y_1)^2$.

$$\begin{aligned} AC^2 &= (3-1)^2 + (7-2)^2 \\ &= 4 + 25 \end{aligned}$$

so AC is $\sqrt{29}$.

Special collections of pairs and points

The distance formula can be used as the starting point for an important development. Think of a general point, P, with coordinates (x, y). The square of the distance of P from the point $(1, 2)$, which we can call A, can be obtained directly from our formula: it is

$$(x-1)^2 + (y-2)^2.$$

I said that P is a general point; that is to say it is free to roam anywhere in the plane – in other words, x and y can take any values. Suppose we restrict P by demanding that its distance from A is fixed, equal to 3, say. Then P would be constrained to lie in a circle with centre at A and radius 3. This is a *geometric* constraint on P, but it has an *algebraic* equivalent:

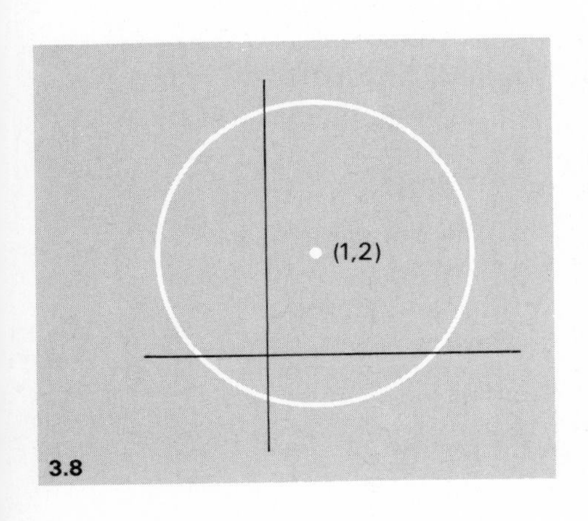

3.8

$$(x-1)^2 + (y-2)^2 = 9.$$

This is interesting enough as it stands, but just think of the implications. A purely geometric notion, a circle, has been described in a purely algebraic way. It is in this way that the

Equation of a circle centre (a, b) and radius r is
$(x-a)^2+(y-b)^2=r^2$

possibility is presented of tackling geometric problems by the methods of algebra.

Think about this the other way round. What does the algebraic condition

$$(x-1)^2+(y-2)^2=9$$

actually mean? It means that the point with coordinates (x, y) is constrained in such a way that the value of

$$(x-1)^2+(y-2)^2$$

must be 9.

For example (2, 3) will not do, because the expression comes to

$$(2-1)^2+(3-2)^2$$

which is 1; in particular, it is not 9. On the other hand, the point (4, 2) does satisfy the condition, because

$$(4-1)^2+(2-2)^2$$

is 3^2, which is 9. If you were to plot every possible point that satisfies the condition you would get a circle. But without actually plotting the points you can predict that the condition describes a circle, because Pythagoras' theorem tells us that $(x-1)^2+(y-2)^2$ is the square of the distance of (x, y) from (1, 2). The condition thus tells us that this distance is fixed – so the point is constrained to lie on a circle. The circle is one kind of curve but before we go on to describe some other curves, let's look at this idea, going from the algebraic context to the geometric situation, in a more general context.

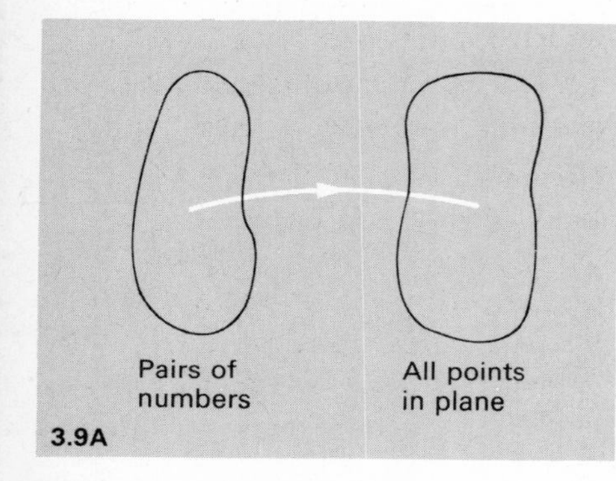

3.9A

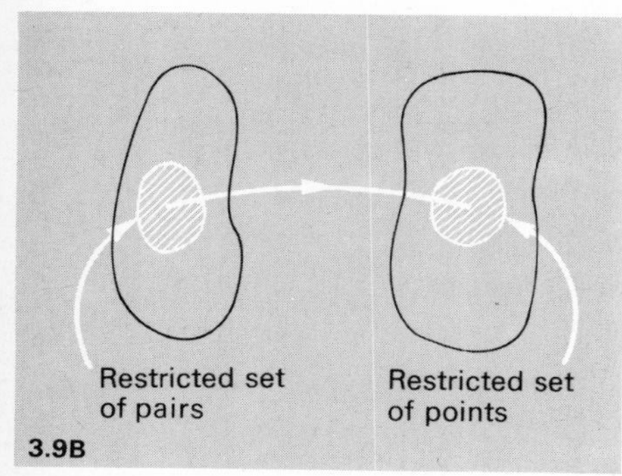

3.9B

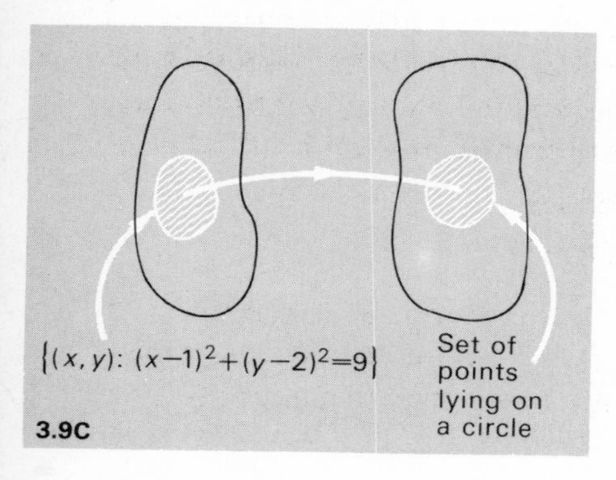

3.9C

We have already noted that the set of all pairs of numbers (x, y) can be pictured geometrically as the set of all points in the plane. If no restrictions are imposed on the values of x and y, then there is no restriction on the points; we get all the points. If x and y are required to satisfy some sort of restriction, however, the possible pairs (x, y) will be restricted and so will the points in the plane. We will get only a *subset* of all the possible pairs, and so just a subset of all the possible points. For example, we have just seen that if x and y are restricted to those values for which

$$(x-1)^2+(y-2)^2=9,$$

then we get a particularly nice subset of points, the points on a circle.

It is convenient to introduce a little notation here. We write the set of pairs as

$$\{(x, y) : (x-1)^2+(y-2)^2 = 9\},$$

which is read 'the set of pairs (x, y) such that $(x-1)^2+(y-2)^2$ is equal to 9'.

More curves

Another curve which can be described algebraically using our distance formula is the parabola. A parabola looks like this: it is the path traced out by a point which is constrained to move so that its distance from a fixed point is not constant (as with the circle), but equal to its distance from a fixed line. The algebraic analogue of the parabola takes an especially useful form if the fixed line is taken parallel to the x-axis. Suppose we take the line through the point with coordinates $(0, -h)$ and let us use (a, b) for the coordinates of the fixed point. There is no great virtue in these choices except that this choice of line gives a convenient form for the algebraic description.

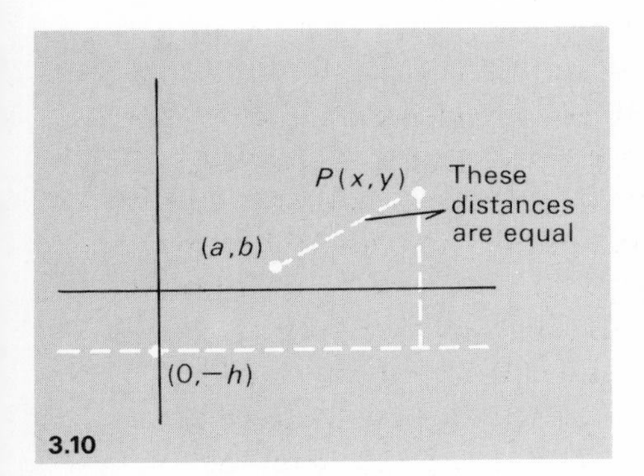

3.10

Look at Fig. 3.10. What is the distance of P from the fixed line? It is just $y+h$. And the distance of P from (a, b) can be worked out from our distance formula: its square is $(x-a)^2+(y-b)^2$. So the geometric condition can be expressed algebraically as

$$(y+h)^2 = (x-a)^2+(y-b)^2.$$

What does this mean? Let us just translate back to 'physical' language so that we know exactly where we are. If you were to plot all the points whose coordinates satisfied this condition, they would form the parabola.

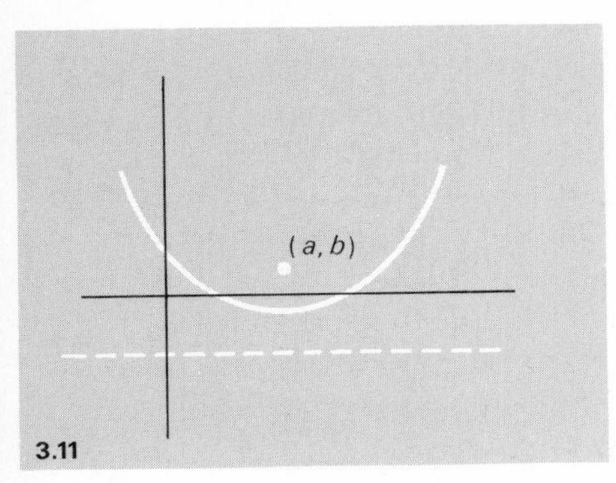

3.11

This can be written in a different form by using the formula for multiplying out expressions like $(\ )^2$ that we worked out in the previous chapter. It turns out to be equivalent to

$$y = \frac{x^2}{2h+2b} - \frac{a}{h+b}x + \frac{a^2+b^2}{2h+2b}$$

Remember that $(x+y)^2 = x^2+2xy+y^2$ —see p.

Don't worry about the details – remember that in any particular case, a, b, and h would be numbers and the whole expression would look far less awe-inspiring: it might be something like $y = x^2-2x+3$, for example.

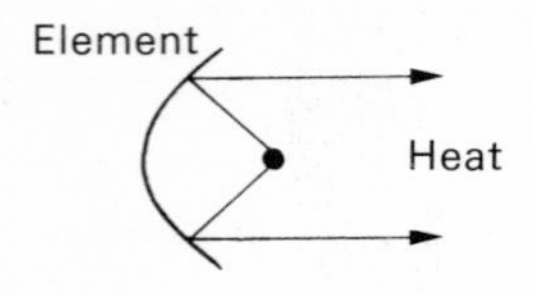

Parabolas: if a mirror is made in the shape of a parabola, light from the fixed point, (*a*, *b*) in the text, is reflected parallel to the axis of the parabola–that is why car headlamps and electric fires are shaped like parabolas.

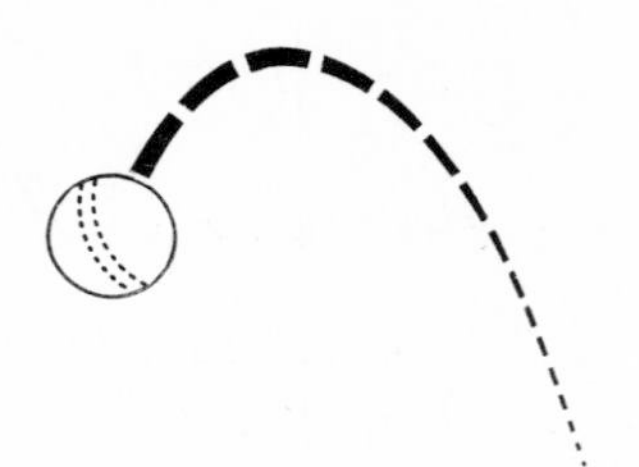

A cricket ball thrown in the air follows the shape of a parabola–if there is no air resistance!

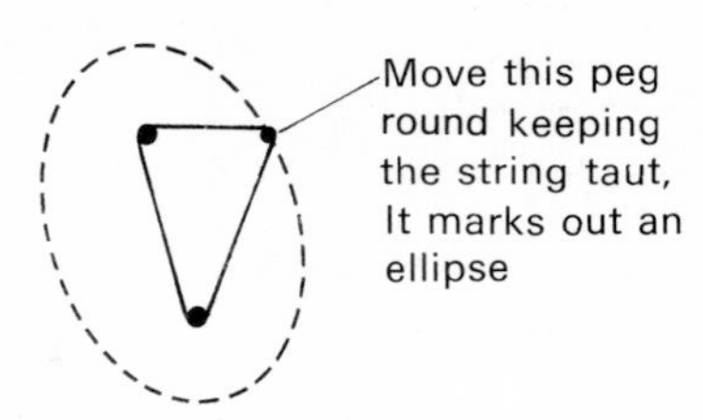

If you want to dig a nice elliptical rose-bed all you need to mark it out is a piece of string and three pegs.

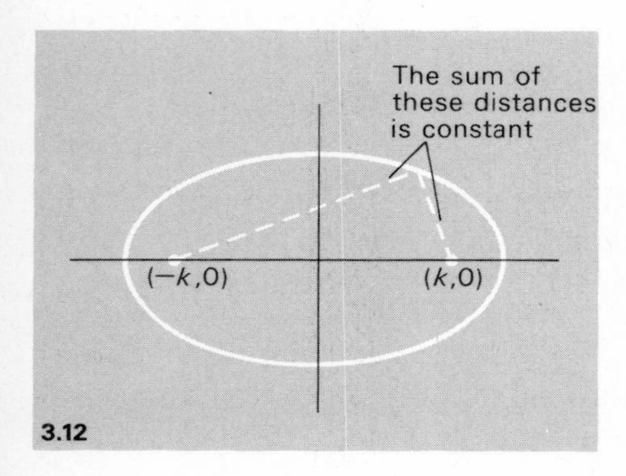

3.12

A curve related to both the circle and the parabola is an ellipse. This time, the condition satisfied by any point, P, on the curve is that the *sum* of the distances of P from *two fixed points* is constant. Again, the algebraic specification of the subset comes from the basic distance formula. Let's take the fixed points as $(-k, 0)$ and $(0, k)$. There is nothing special about these symbols – but choosing points like these, either side of the origin, gives a nice algebraic form of the condition. If we take $2a$ as the sum of the distances, the condition to be satisfied by x and y turns out to be

$$\frac{x^2}{a^2} + \frac{y^2}{b^2} = 1,$$

where $b^2 = a^2 - k^2$. (Variations in the values of a have the effect of making the ellipse fatter or thinner and the smaller the value of k, the closer together the fixed points become and the more nearly does the ellipse approach a circular shape – if they coincide, the curve becomes the path traced out by a point at a constant distance from *one* point, so it *is* a circle. In fact you can see this directly from the algebra. If you put $k = 0$, then, because $b^2 = a^2 - k^2$, the condition turns out to be

$$\frac{x^2}{a^2} + \frac{y^2}{a^2} = 1,$$

in other words, $x^2 + y^2 = a^2$, which is the equation of a circle.)

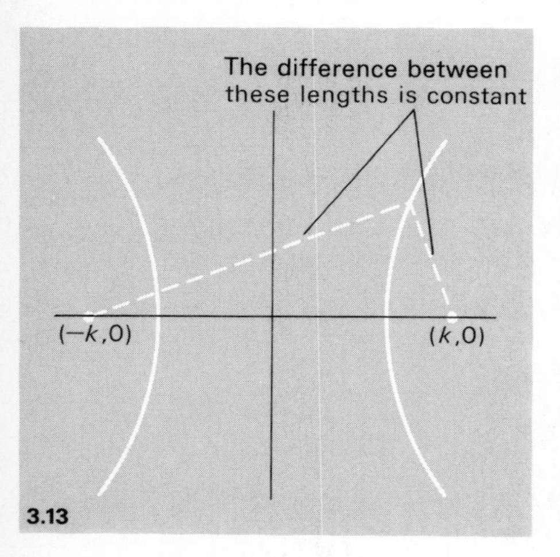

3.13

Another related curve is the *hyperbola*: this is the curve traced out by a point moving so that the *difference* of its

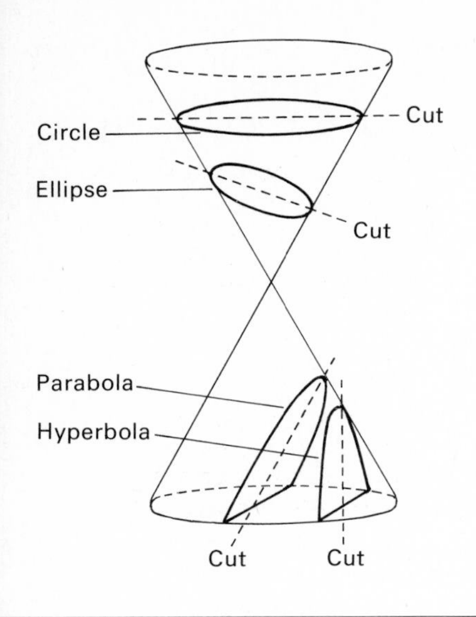

Conic sections: if you slice through a cone and look at the face of the cut surface, its shape will be a circle or an ellipse or a parabola or a hyperbola–that is why the curves are called conic sections.

distances from two fixed points is constant. This time the algebraic condition works out to be

$$\frac{x^2}{a^2} - \frac{y^2}{b^2} = 1,$$

where $2a$ is the constant difference and again $a^2 - b^2 = k^2$.

There are, of course, many many other curves with considerable geometric interest and many equations with illuminating geometric curves. We have chosen these particular examples because they form a related collection which in any case are amongst the most important curves. They should give you an idea of the way that geometric properties can be converted into algebraic conditions, if not the way that geometric *problems* can be tackled algebraically. We shall be seeing a little of this from time to time later, but as a subject in itself it is worth many books in its own right. We have to resist the temptation to pursue it here; make a promise to yourself to take it up sometime. For the present, all we really need is the idea of a curve described by an algebraic condition imposed on the coordinates of a general point on the curve. The algebraic specification of a curve is called the *equation* of the curve.

Regions of the plane

Other types of subset of the points of a plane can be described by algebraic conditions of a different kind. For example, if x is restricted to be positive, the set of points constitutes a half-plane.

If y is also restricted to be positive, we get a 'quadrant' of the plane.

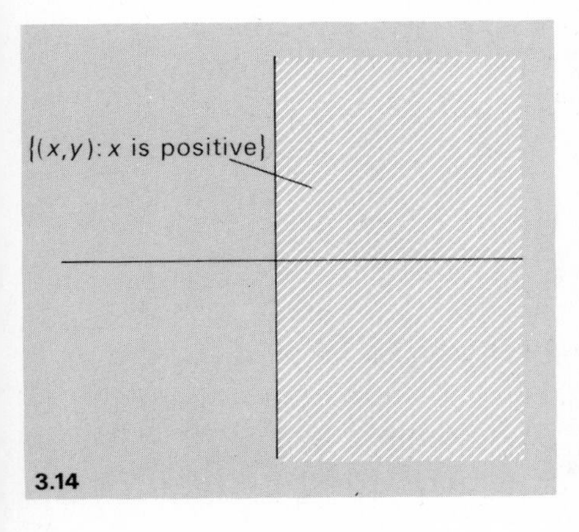

3.14

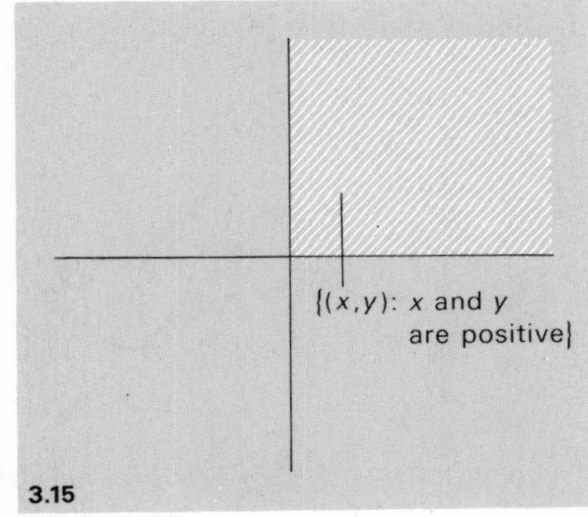

3.15

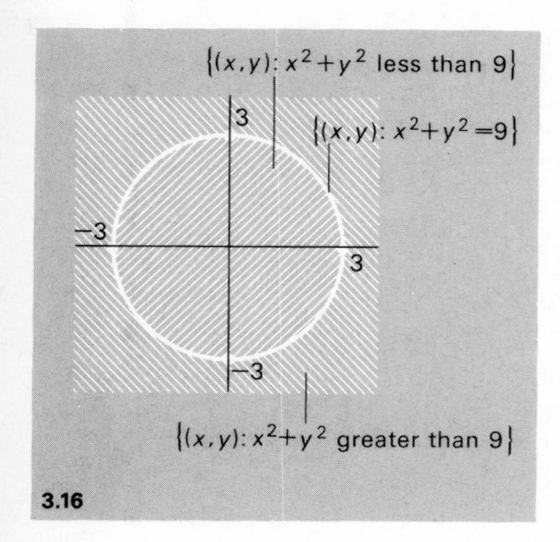

3.16

So it is possible to use algebraic conditions to describe regions as well as curves. This is hardly surprising, because curves themselves specify regions of the plane. For example, a circle separates the points of the plane into three subsets – the points outside the circle, the points inside the circle and the points on the circle. Each of the regions can be described by conditions to be satisfied by x and y for the point (x, y) to belong to the region.

Straight lines

It would be rather perverse to leave this set of examples without seeing an algebraic specification of perhaps the simplest of all subsets of points – a straight line.

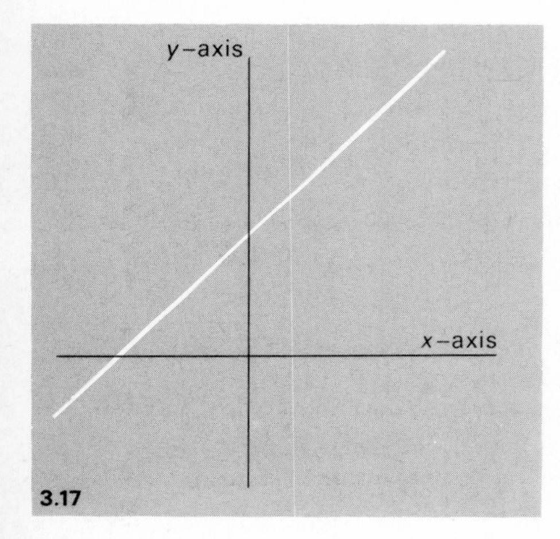

3.17

Just as with the other curves, the task is to express algebraically the essential feature that makes a straight line a straight line. A moment's thought might convince you that this is not as easy as might at first appear. With the other curves we had a geometric specification in terms of distance that we were able to translate directly into algebra. Strangely enough it is made easier when we introduce a coordinate ('map reference') system, for then a line existing in limbo becomes in a sense anchored in the plane by the coordinate system. Of course, this is entirely deceptive because the coordinate system can go anywhere! But that is beside the point – the algebra is just a device to model the geometry and we can choose whichever model we wish.

Having introduced a coordinate system we can characterize the line by the property that in moving from one point on the line to another, the increments in the y-direction and the x-direction are always in the same ratio. Thus

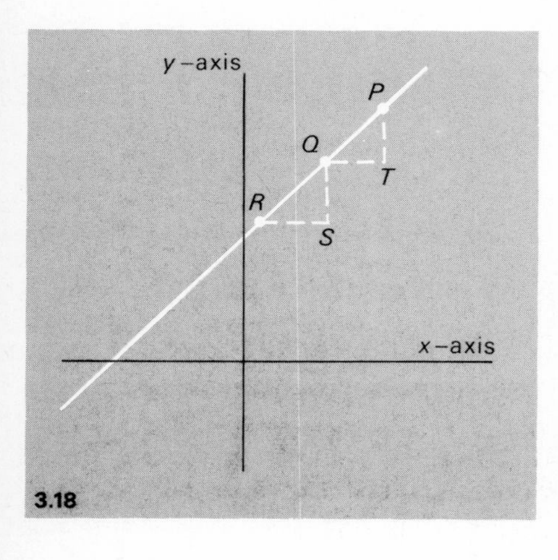

3.18

$$\frac{PT}{QT} = \frac{QS}{RS}$$

whatever points P, Q, and R we choose on the line. One way to obtain the equation of the line is to make cunning choices for R and Q: choose them to be the points where the line intersects the axes. Suppose these are the points $(a, 0)$ and $(0, b)$. In diagram 3.19, a is negative; that is why the distance RO is labelled $-a$, to give a positive number, which we need in order to represent a length – for example if a were -2, the length RO would be 2, which is $-(-2)$. Let the coordinates

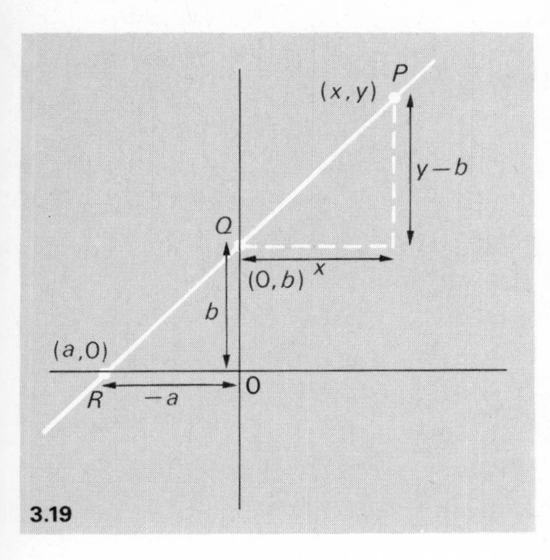

3.19

Equation of a straight line: if x and y are related by a condition of the form $ay+bx=c$ then the point (x,y) lies on a straight line. If $a=0$ the line is parallel to the y-axis. If $b=0$ it is parallel to the x-axis. The line intersects the x-axis at $(c/b, 0)$ if b is non-zero and the y-axis at $(0, c/a)$ if a is non-zero.

of P, a general point on the line, be (x, y). Then the characteristic property of the line can be expressed as

$$\frac{y-b}{x} = \frac{b}{-a}.$$

If you juggle this equation around you can get the rather more elegant form

$$ay+bx = ab.$$

Just to give an idea of what is going on, you could try an example, say $y+2x = 2$. Try giving x a few values, working out the value of y and plotting points. You will find that they all lie on a straight line. This method does not work if a is zero, for then $\frac{b}{-a}$ does not make sense. In any case if a is zero, b must also be zero because the line passes through 0. The equation can be worked out in that case using much the same idea: if the line passes through a point (c, d), the equation is $cy = dx$.

There are two other special cases: the method breaks down if the line is parallel to either of the coordinate axes. These are easily dealt with. A line parallel to the x-axis has equation $y = b$, where $(0, b)$ is its intercept on the y-axis. A line parallel to the y-axis has equation $x = a$ where $(a, 0)$ is its intercept on the x-axis.

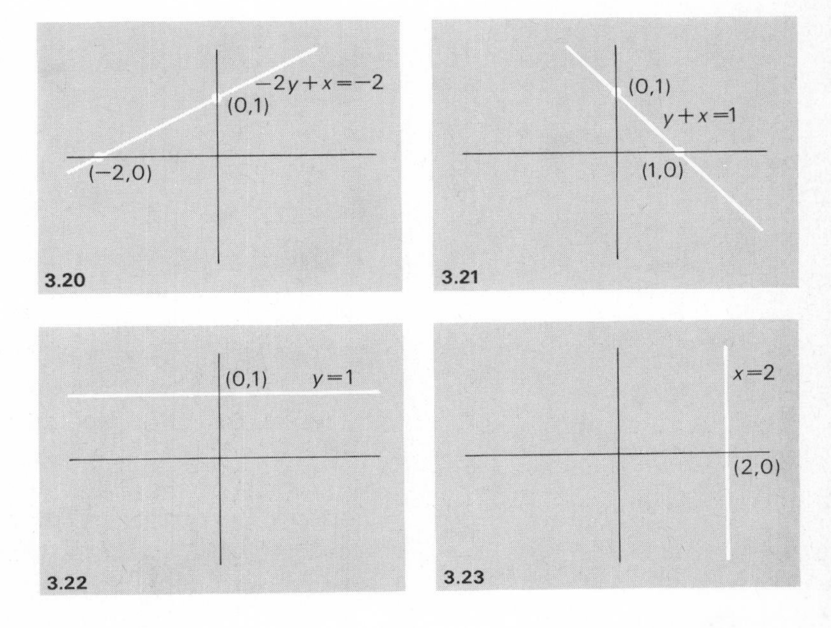

3.20 3.21 3.22 3.23

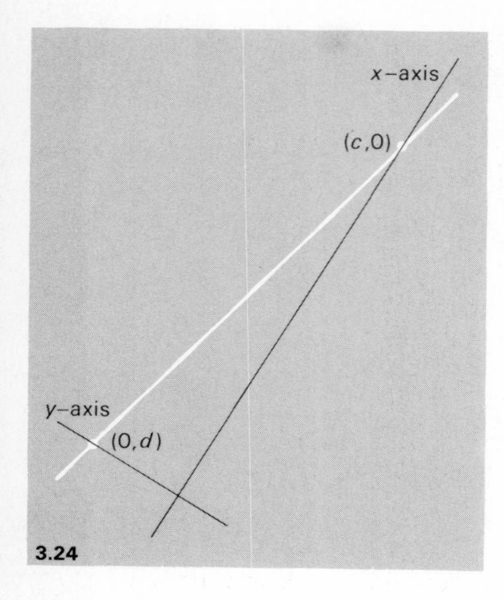

3.24

So that covers all the cases: we now have an equation for every type of straight line. But you may have been worried about what might happen if we had put the axes somewhere else. This raises an interesting point. What could have happened? Only two things could change – the values of a and b. And all that would happen to the equation of the line would be that it would look something like

$$cy + dx = cd\,.$$

I claimed that the effect would raise an interesting point: perhaps you feel that it is about as interesting as a dull thud: the equation looks almost the same as it did before. But that boring observation is just the vital point. *Whenever* you see an equation like

$$cy + dx = cd\,,$$

you *know* it specifies a line. In other words the very *form* of the equation

$$(\text{some number}) \times y + (\text{some number}) \times x = \text{some number}$$

characterizes the 'lineness' of a line. Not only have we modelled a line algebraically, we have captured its essence in the structure of the equation.

This is the sort of thing that excites mathematicians, and they will excite you too once you get a little way into the subject. To a mathematician, an *equation* like this immediately conjures up all the sort of pictorial notions that are usually associated with straight lines – without actually drawing anything: in fact equations of this form are referred to as *linear* equations!

In much the same way, *any* equation of the form

$$(x-a)^2 + (y-b)^2 = r^2$$

specifies a circle, and *any* circle has an equation of this type. And *any* equation of the form

$$\left(\frac{x}{a}\right)^2 + \left(\frac{y}{b}\right)^2 = 1$$

specifies an ellipse, and *any* equation of the form

$$y = ax^2 + bx + c$$

specifies a parabola. Furthermore, with a cunning choice of coordinates *any* ellipse, or parabola can be described by an equation of this standard form.

There is a complete correspondence between each of the curves and each form of equation.

That something as abstract as a jumble of symbols can capture all that the notion of, say, an ellipse entails is an indication of just what a great creative feat of the human mind mathematics really is. This is the stuff of the subject: the search for form, for pattern, and for the unification of ideas.

Functions

One of the most useful tools in the mathematical search for relationships is the concept of function. It has long been implicit in mathematics, but it is only in this century that it has been allowed to play its full part. This flowering was due mainly to the fact that it was at last given a precise definition. Thus it became a respectable entity, it could be used in its own right, rather than in a vague, Svengali role.

The reason why functions occupy such a central position is that *they* are the devices which are used to describe relationships – and relationships is what mathematics is all about. Consider the set of points that constitute a parabola. It can be thought of as specifying a relationship between points on the x-axis and points on the y-axis in the way that the points A and B in the diagram are related. So the number 3 is related to 9, -6 to 36, and so on: to each number is assigned its square. To convey the dynamic nature of this relationship, we use an arrow: we write

$$3 \to 9, \quad -10 \to 100, \; x \to x^2, \; t \to t^2$$

and so on, reading the notation as '3 maps to 9' or 'x maps to x^2' for example.

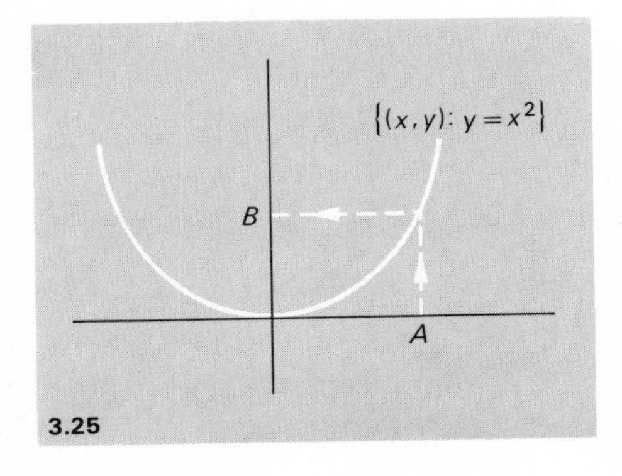

3.25

The rule (in the example 'x maps to x^2') together with the set on which it operates (the set of numbers) and the set it produces (the set of positive numbers) constitutes a *function*.

So a function consists of three ingredients – two sets of objects (more often than not numbers) and a rule.

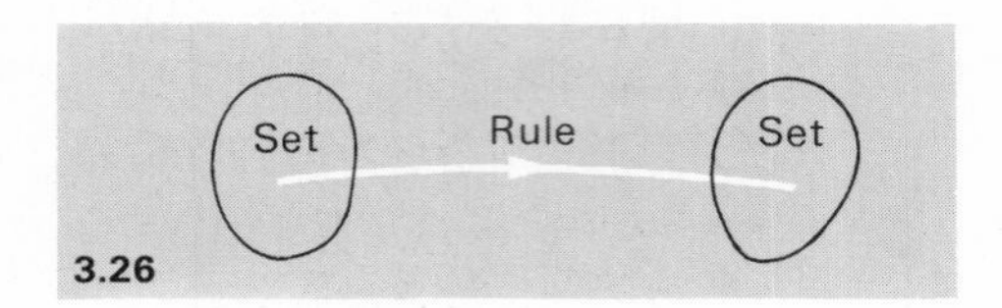

3.26

Objects from one set are processed by the rule to give objects in the other set.

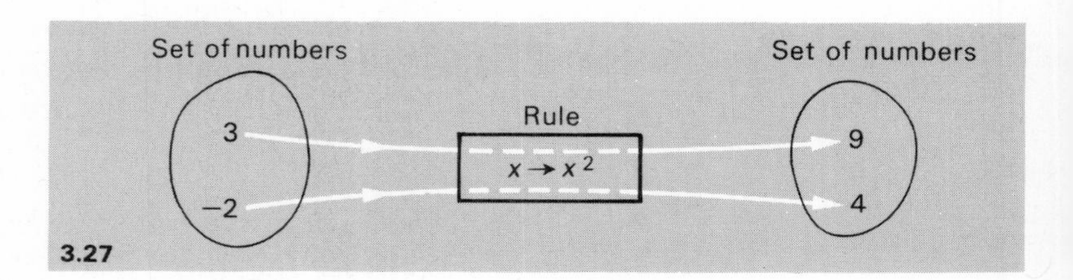

3.27

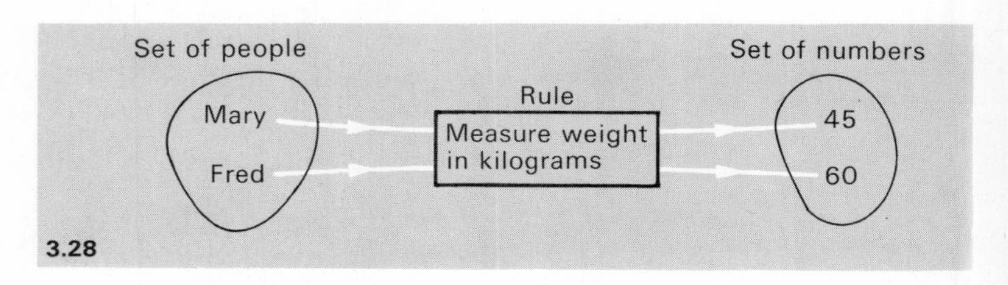

3.28

We shall see more of functions in the next chapter; in this chapter both of the sets of objects will be sets of numbers. There is one more important thing to say now, and that concerns notation. It is often convenient to be able to refer to a function in the course of an argument, and an easy way to do this is to label it with a symbol. For example we could use the symbol f to stand for our 'square' function above and we would then write

$$f: x \to x^2$$

or $f(x) = x^2$. (which is pronounced 'f of x equals x squared')

(Notice that the use of brackets in $f(x)$ is nothing to do with multiplication: it is simply a device to convey the feeling that f in some sense 'processes' the number x.)

As x takes on all its allowable values, the function specifies a set of pairs $\{(x, f(x))\}$. The set of pairs is called the 'graph' of the function. If, as in the present case, these pairs are pairs of numbers, then the graph can be represented as a subset (specific choice of points) of the plane – this is the more traditional meaning of the now more general term 'graph'. Don't worry too much about these more general ideas of graphs and functions, we shall be coming back to them in the next chapter. For the moment, the only functions we are interested in are those where an algebraic formula acts as a rule connecting together numbers.

Optimization problems

In practical applications of mathematics, situations often arise where an optimum choice has to be made from a set of allowable possibilities – maximizing power output minimizing energy consumption, maximizing profit are all fairly obvious examples. If mathematical methods are to be used to tackle such a problem, at least two stages have to be tackled. First, the problem has to be formulated in mathematical terms – this requires a mathematical description of the quantity to be optimized and a mathematical description of the set of possible situations. Second, a mathematical technique has to be applied which will enable the optimum choice to be made.

To show you the idea, here is a rather artificial problem that does not really need a mathematical approach but which nevertheless illustrates the ideas. Suppose a self-sufficiency enthusiast wants to enclose a rectangular chicken run with a roll of wire 30 metres in length. What length should the sides be so as to enclose the maximum area? You can probably guess the answer: we can confirm it like this – and our method will be applicable to less obvious problems.

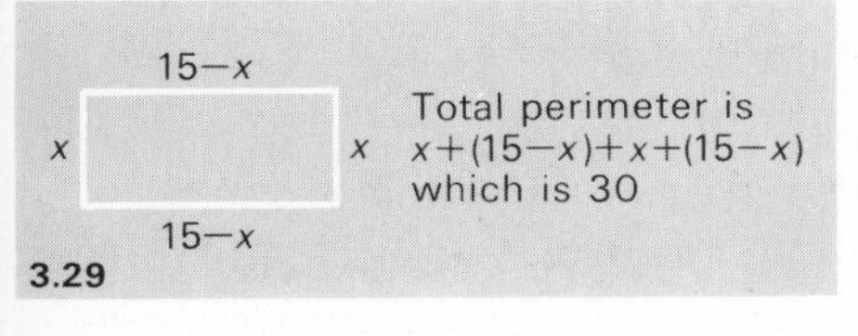

3.29

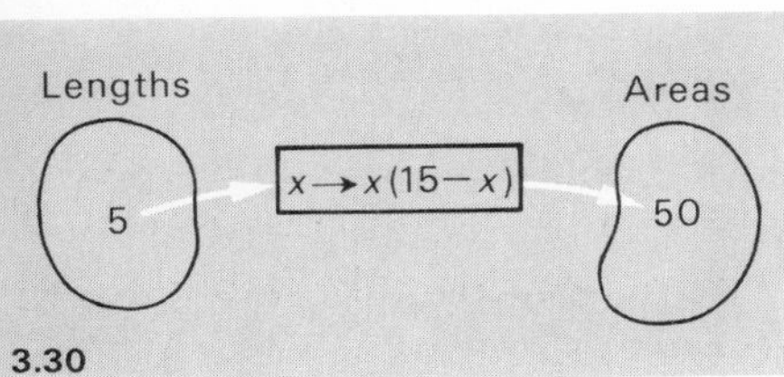

3.30

To express the situation in mathematical terms we can introduce a symbol, let's choose x, to stand for the length of one side. We do not need another symbol to stand for the other side's length: we can use the fact that the total perimeter is 30 metres. The other length must be $15-x$; so the area is $x(15-x)$.

This gives us the rule for the function, f, which converts the length, x, into area:

$$f(x) = x(15-x)$$

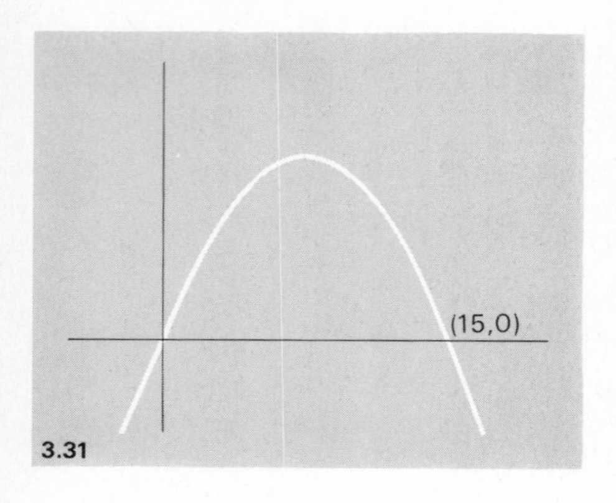

3.31

If the brackets on the right-hand side are removed, we get $15x - x^2$ and if you look back to p. 61 you can recognize that the graph of f is a parabola. It looks like this.

The minus sign with the x^2 shows us that it is 'upside down' – as x increases, y is large and negative. (Try it with $x = 100$ and $x = -100$.) The earlier form, $x(15 - x)$ shows us that $f(x)$ is zero when x is zero and when x is 15.

How miraculous that a problem set out initially in terms of a rectangle generates a *curve*! Remember, though, that this curve is not telling us directly the shape of the field – we're not going to have to build a parabolic fence! But it contains the information which will tell us how long the sides of the rectangular field will have to be. Our problem is to find out how to extract that information.

Looking at the graph, we can see that the largest area arises at the 'hump' – this corresponds to the value of $7\frac{1}{2}$ for x, giving sides of $7\frac{1}{2}$ and $15 - 7\frac{1}{2}$ which is also $7\frac{1}{2}$; the optimum shape is a square, with area $56\frac{1}{4}$ square metres.

Now let's think a little. The parabola we have drawn is the graph of the function f for which $f(x)$ is the area of the rectangle when one of the sides is x. Clearly, the values that x can take are restricted. It certainly cannot be greater than 30, for that is the total length of netting available. In fact it must be restricted to values between 0 and 15 – this is obvious from the graph (values of x outside this range would give a negative area) or by thinking directly about the problem. The point I want to make here is that in setting up any mathematical model, the interface between the physical situation and the mathematics is a treacherous no-man's land. Here, for example, the function

$$f: x \to x(15 - x)$$

models the chicken run for only a limited range of values of x; without having that limitation clear in our minds we might conclude that $f(x)$ could take some optimum value which was actually inadmissable. You could imagine the disastrous consequences that might arise in a highly complicated model of, say, a blast furnace, if an arrangement were suggested that led to a setting outside the recommended operational limits.

A second point that arises from this problem is rather different. Did you guess the answer to this problem at the beginning? If you did, *why* did you guess it? Your response is probably 'what else could it be?' Although that may appear to be a rather aggressive response to a somewhat stupid question it nevertheless shows some sound mathematical feeling inside you somewhere. You must have reasoned, maybe subconciously, that the sides of the rectangle play a symmetric role – whatever can be said about one of the sides applies just as well to the other. And if no one side can claim a special status, the conclusion that all sides must be equal is compelling. That is just the sort of reasoning that mathematicians look for in more complicated problems. To mould it into a rigorous proof is not easy, but as I have tried to show you earlier, mathematics is not all cold hard facts and straightforward deductive reasoning: it has more than its fair share of inspired guesswork and creative insight.

The final point about this problem is of a more practical nature: the same principles that we adopted can be employed in a less obvious problem. If the possible choices open to us can be regarded as the input and the values of the quantity to be optimized as the output, then the task of optimization amounts to looking for the highest point on the graph of the function.

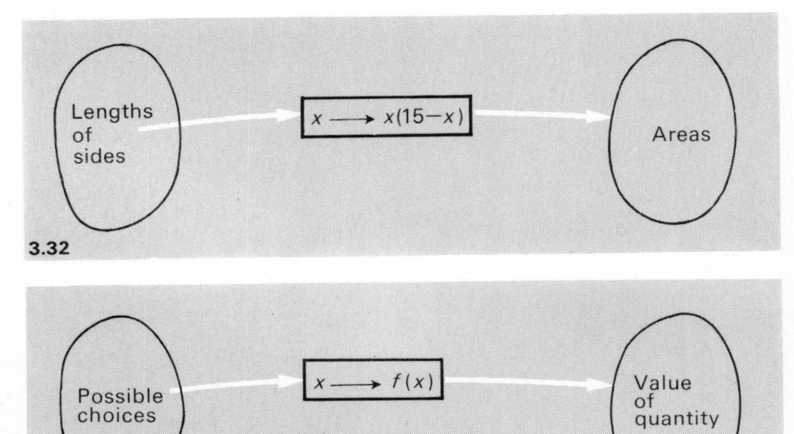

3.32

3.33

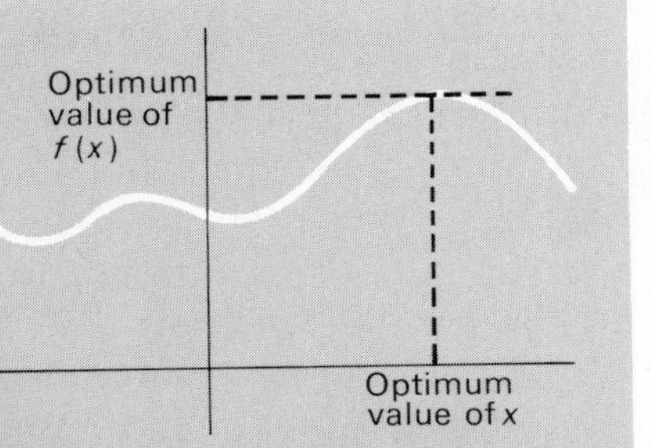

3.34

In our particular example, it was easy to spot this point. In more complicated cases it is not so easy, and special techniques have been developed to find such points without needing to sketch the graph. We'll look at them in Chapter 7.

How publishers can try to save money

Here is an example of a more difficult problem to show you what I mean – we shall finish it off in Chapter 7 when we have developed sufficient mathematics. A publisher wants to decide which size of page he should use for maximum economy subject to the following constraints. The page is to have an area of 300 square centimetres, there is to be a $2\frac{1}{2}$-centimetre margin at the top and bottom and a 3-centimetre margin each side (or for awkward authors 2 centimetres on one side and 4 on the other, it does not make any difference to the problem). His problem is to decide the dimensions of the page which gives the maximum amount of print on each page. (When we have solved this one, you will be in a good position to judge the aesthetes from the others in the publishing world!)

Remember that
$f: x \to x^2+2$
is another way of writing
$f(x) = x^2+2$
It is read 'f maps x to x^2+2

If the publisher has been fortunate enough to have read this book, he might well set up a function which describes the relationship between the length of side and the print area. As usual, to cope with all possible situations at once, we use a symbol for the length of side. If the side has length x centimetres, than the other side must have length $\dfrac{300}{x}$ centimetres, because when you multiply these two lengths the result is a page area of 300 square centimetres. The rectangle of print would then have sides of lengths $(x-6)$ centimetres and $\left(\dfrac{300}{x} - 4\right)$ centimetres and so its area would be $(x-6)\left(\dfrac{300}{x} - 4\right)$ square centimetres.

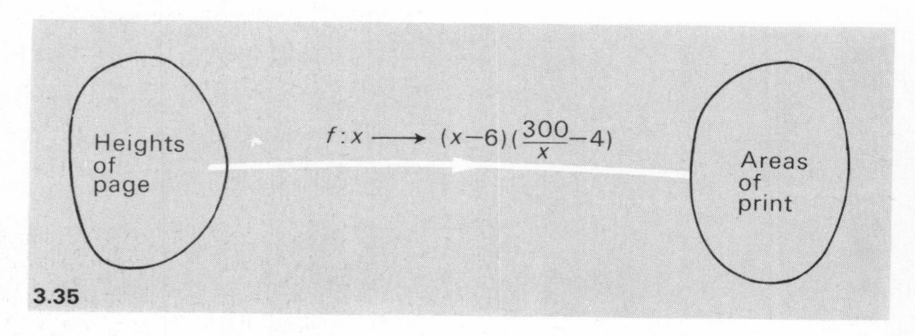

3.35

This time, we have to look at the graph of the function

$$f: x \to (x-6)\left(\frac{300}{x} - 4\right)$$

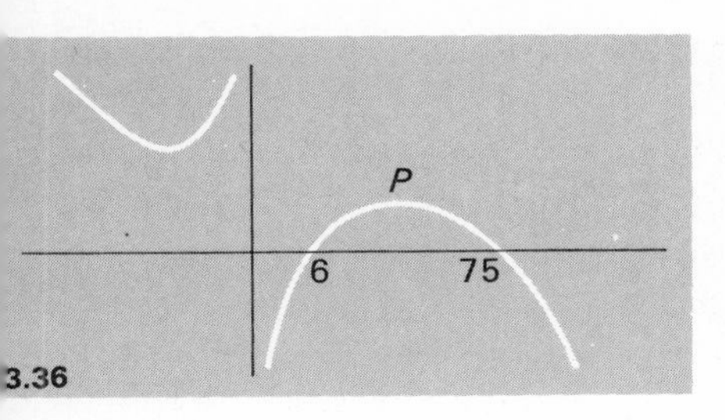

3.36

and it is not so easy to sketch the graph or to spot the optimum point. (By the way, it is fairly clear that, because the area has to be positive, x has to be greater than 6 and $\frac{300}{x}$ has to be greater than 4 – you would expect this from the original statement of the problem: why?) In fact, the graph turns out to be like Fig. 3.36. The best choice for the publisher is the value of x at the point P – we shall work it out in Chapter 7. If you want to know what the answer is you can plot the graph and estimate the value of x. But it is possible to *calculate* it precisely, once we have available a special technique. This is something we shall tackle in Chapter 7.

How to deliver furniture

A manufacturer of home-assembly furniture has undertaken to supply at least 100 hi-fi units and at least 150 bedside cupboards each week to a warehouse. Because of the limitations on administration, warehouse space, and transport, the warehouse cannot cope with more than 600 items each week, nor can it accept a total weight in excess of 9600 kilograms. The hi-fi units weigh 20 kilograms each and the bedside cupboards weigh 12 kilograms each. The manufacturer's problem is this: he makes £4 profit on every hi-fi unit and £3 profit on every bedside cupboard; how should he make up his delivery to secure maximum profit?

He might reason like this. Since he makes £4 on each hi-fi unit, it would make sense to send as many of these as possible. The most he can send is 480, because of the limitation on weight. But this would contravene his agreement to supply 150 bedside cupboards, so perhaps he should send 150 cupboards (weighing 1800 kilograms) and make up the remaining weight of 7800 kilograms with 390 hi-fi units. This would give a profit of £$(390 \times 4 + 150 \times 3)$, which is £2010.

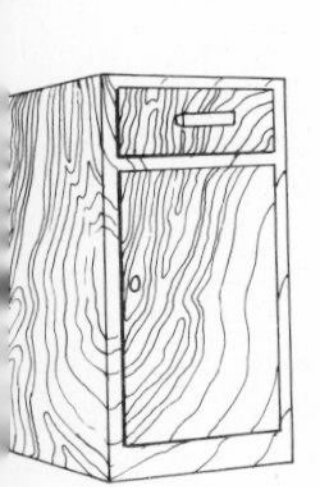

Cupboards
weigh 12 kg
make profit £3

Hi-fi's
weigh 20 kg
make profit £4

On the other hand, he might realize that, although he makes less profit on bedside cupboards, he can send a lot more, because they are lighter and, since he is sending only 440 items under his original arrangement there may be more scope for profit. He may well experiment with various combinations, but he could adopt a more systematic approach.

What does he want to know?

The optimum number of hi-fi units and bedside cupboards.

So let this optimum number of hi-fi units be x and the optimum number of bedside units by y.

What possible values can x and y take?

To answer that, we have to express the supply constraints in terms of x and y.

1. He must supply at least 100 hi-fi units. So x must be greater than (or equal to) 100.
2. He must supply at least 150 cupboards. So y must be greater than (or equal to) 150.
3. The total number of items must not exceed 600. So $x+y$ must be less than or equal to 600.
4. The total weight must not exceed 9600 kilograms. The hi-fi units weigh 20 kilograms each; x of them would weigh $20x$ kilograms. The cupboards weigh 12 kilograms each; y of them would weigh $12y$ kilograms. The weight constraint means that

   $20x + 12y$ must be less than or equal to 9600.

To see how to proceed, it will be useful to illustrate these constraints graphically. But, first, it is convenient to introduce a little shorthand. Just as the symbol '=' is used to stand for 'is equal to' we have symbols to stand for 'is less than or equal to' and 'is greater than or equal to'. We write

$\leqslant$ for 'is less than or equal to'

$\geqslant$ for 'is greater than or equal to'.

So our constraints on x and y now look like this:

1. $x \geqslant 100$
2. $y \geqslant 150$
3. $x+y \leqslant 600$
4. $20x+12y \leqslant 9600$

How do we represent these on a diagram?

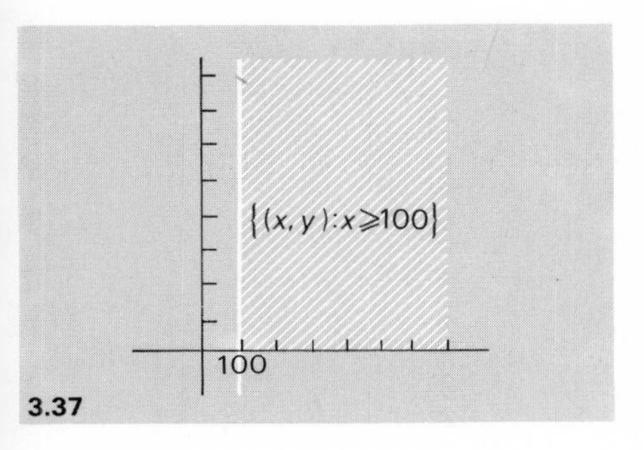

3.37

Each pair of values of x and y – each of the transport configurations – can be represented by a point in the plane. If there were *no* constraints on the manufacturer, he could select *any* point in the plane as representing a possible arrangement. But each constraint will correspond to restricting the possible points to a subset of the plane. For example, $x \geqslant 100$ must mean that the point has to be to the right of the line $x = 100$. $y \geqslant 150$ means that the point must be above the line $y = 150$.

For both conditions to be satisfied, the point must be chosen to lie in *both* of these regions.

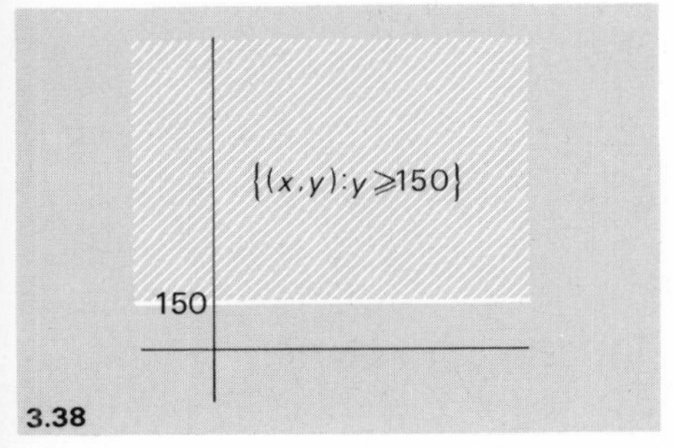

3.38

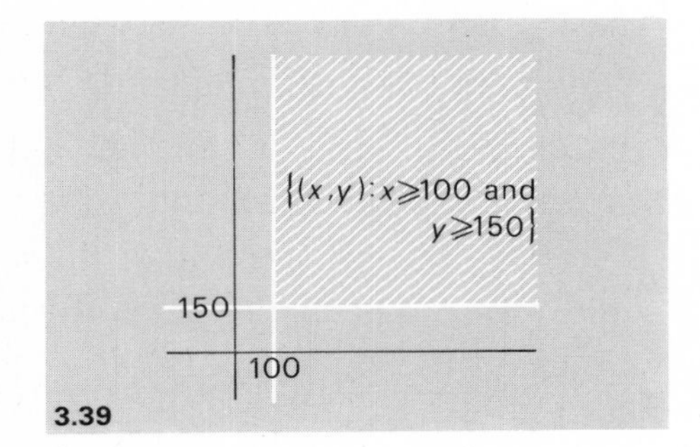

3.39

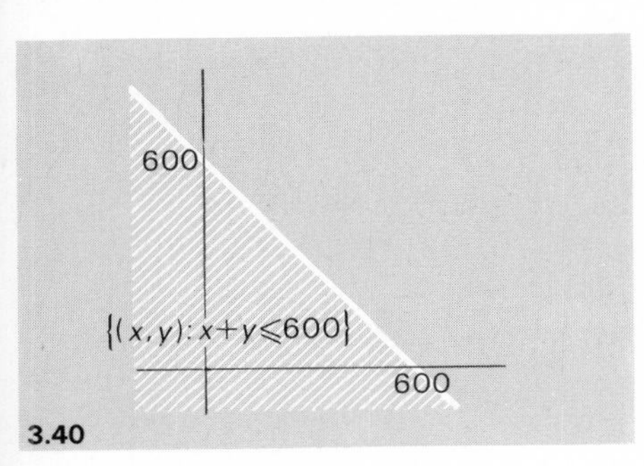

3.40

The other two constraints can be dealt with in much the same way, but with a little more technical difficulty. Just as the line $x = 100$ divides the plane into two regions, with $x \geqslant 100$ in one of them and $x \leqslant 100$ in the other (and $x = 100$ on the common boundary), so the line $x+y = 600$ has the same effect. On one side of the line $x+y$ is greater than 600. On the other side it is less than 600.

In the same way, to locate the region in which $20x+12y \leqslant 9600$, we draw the line $20x+12y = 9600$. Putting $x = 0$ gives $y = 800$ and so $(0, 800)$ is one point on the line. Putting $y = 0$ gives $x = 480$ so $(480, 0)$ is on the line. The line must be the one through these two points. We are now in a position to locate the region representing possible pairs of values of x and y. It is the set of points common to *all* of the regions we have drawn. It is bounded by the polygon $ABCD$. There is a minor problem – it does not make sense to send $125\frac{1}{2}$ hi-fi units. Only whole numbers are acceptable for x and y, so really we need to be looking at a 'lattice' of points inside $ABCD$. To a large extent, this is a technical difficulty so we need not worry about it just yet.

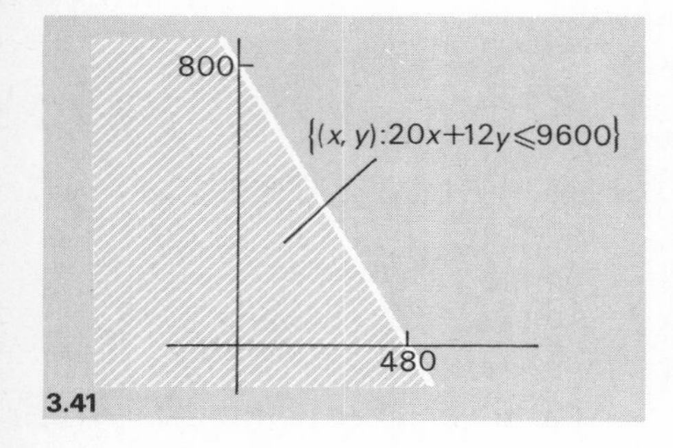

3.41

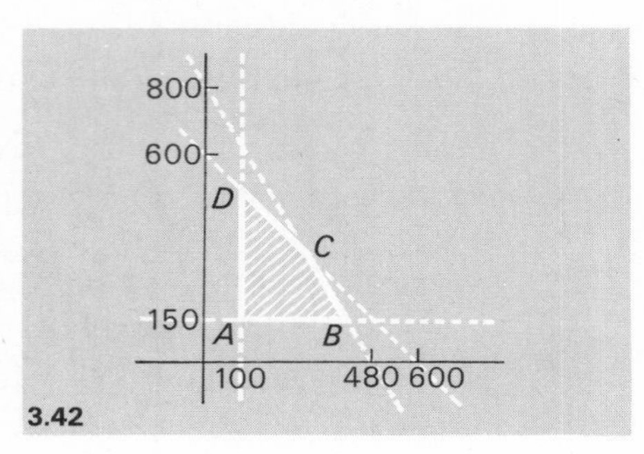

3.42

We have represented the set of all possibilities from which the manufacturer can select his shipping arrangements. We now need a mathematical device for selecting the optimum. What does he want to optimize? His profit. What is the profit in terms of x and y? He makes £4 on each hi-fi unit and £3 on each cupboard. So a shipping of x units and y cupboards yields

$$£(4x+3y) \text{ profit}.$$

The next step is rather cunning. Instead of asking directly how to maximize the profit we first note a point that will prove very useful. When x and y are both 100, for example, the profit is £700. The profit is also £700 for every other point on the line $4x+3y = 700$ – for this equation actually says 'profit equals 700'.

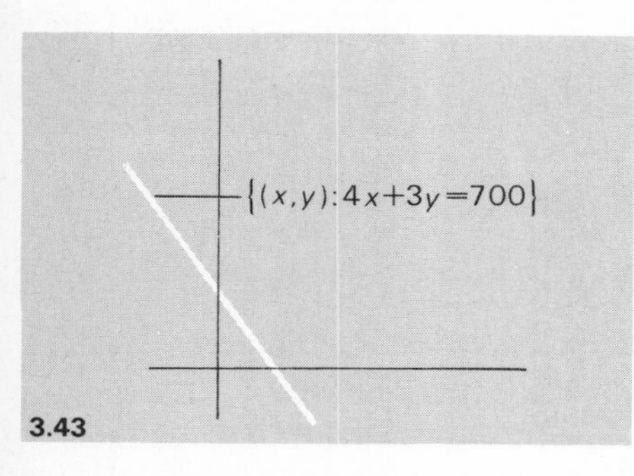

3.43

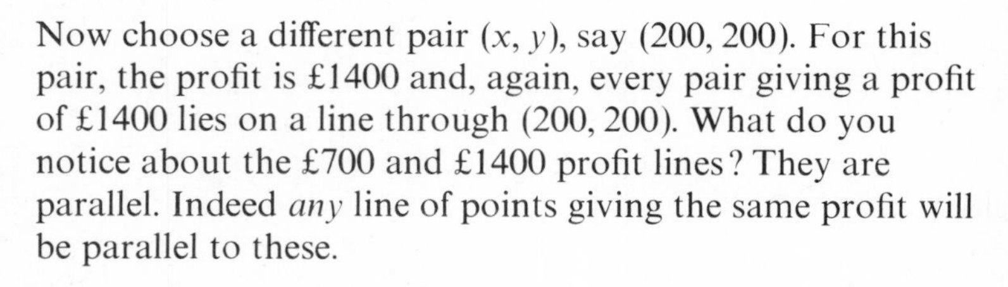

Now choose a different pair (x, y), say (200, 200). For this pair, the profit is £1400 and, again, every pair giving a profit of £1400 lies on a line through (200, 200). What do you notice about the £700 and £1400 profit lines? They are parallel. Indeed *any* line of points giving the same profit will be parallel to these.

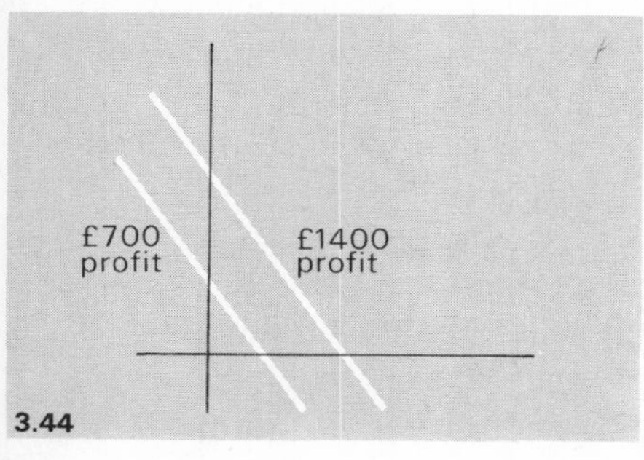

3.44

The next step is to combine the two ideas – the set of parallel profit lines and the set of allowable pairs (x, y).

The dotted line and the lines made up of long dashes are sample profit lines. But no point of the dashed line belongs to

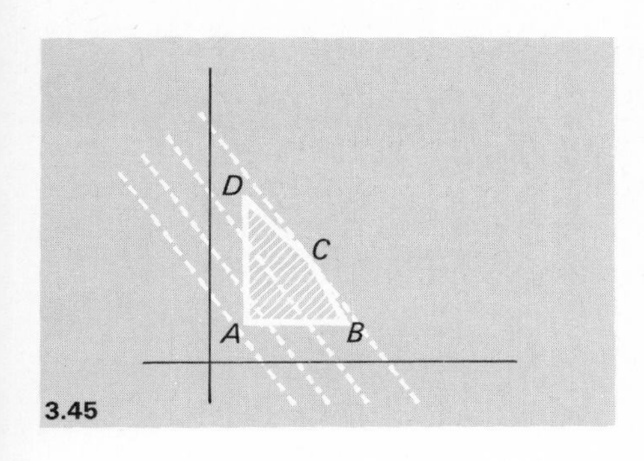

3.45

the shaded region of allowable points. So whatever profit is represented by each of the dashed lines, it is not possible to make that profit. The dotted line profit can be made by choosing any of the points lying both on the line and in the region.

How do we choose the point in the region that gives the maximum profit?

If you look at the £700 and £1400 profit lines that we drew above you will see that the £1400 line is further from 0. This is no coincidence, if you think about it for a few moments, you will soon convince yourself that the further from 0 the profit line lies, the larger the profit it represents. So, to find the maximum possible profit, we need to choose a profit line which is as far as possible from 0 but which still contains at least one point in the region of allowable points.

Can you see that the maximum possible profit must occur at one of the vertices A, B, C, or D? It is fairly obvious that it does not occur at A, but the choice between B, C, and D is not so straightforward – it depends on the slope of the profit lines.

But having used the geometry to give us insight into the problem, we can now go back to algebra. We *could* keep drawing parallel profit lines further and further from 0 and stop when we got just to an extremity of the region, but since we *know* the maximum profit will be at B, C, or D, we need only calculate the profit at each of these points and choose the best. We have reduced the choice of possibilities to just three. First we have to calculate the coordinates of the points B, C, and D. We can do this as follows.

The point D is on the line

$$x = 100$$

and on the line

$$x + y = 600$$

so the values of x and y at D must satisfy each equation. This means that x must be equal to 100 (because of the first equation) and the second equation thus tells us that

$$100 + y = 600$$

so

$$y = 500$$

You can work out the coordinates of B and C in a similar way. The upshot is that

at D, $x = 100$, $y = 500$

at B, $x = 390$, $y = 150$

at C, $x = 300$, $y = 300$.

Now we can work out the profits.

At D the profit is £$(4 \times 100 + 3 \times 500)$, which is £1900.

At C the profit is £$(4 \times 300 + 3 \times 300)$, which is £2100.

At B the profit is £$(4 \times 390 + 3 \times 150)$, which is £2010.

It is therefore possible to say with certainty that the maximum possible profit is £2100 and that this is achieved by sending 300 hi-fi units and 300 cupboards.

A number of points arise from this example that are worth spending a little time on. In the first place, this particular example may seem too simple to need the type of technical armoury that we have brought to bear on it. In a sense this is true, but even if you get the same answer by trial and error would you have been so convinced that it *does* give the maximum profit? Anyway, the method we have used is not really as bad as it may seem – you have had to cope with an explanation of the method at the same time as solving the particular problem. The actual technique does not involve an awful lot more than drawing straight lines and finding the coordinates of the points where lines intersect. Another thing to note is that if the manufacturer's profit changes – say he found that he could improve his profit on cupboards – he has only to recalculate the profit at the vertices to find the new optimum arrangement.

The latter point has some relevance to other types of problems where this technique has application, situations where most of the parameters are constant, with just one aspect varying. For example, a chicken farmer has to mix up feedstuffs to make sure that they contain a minimum of vitamins, minerals, and so on. The amount of, say, iron in a particular foodstuff is proportional to the amount of a particular food – double the quantity and you double the amount of iron from that food. The total iron ration is made up of contributions from various foodstuffs. The constraints appear in a linear way, just as in our furniture example.

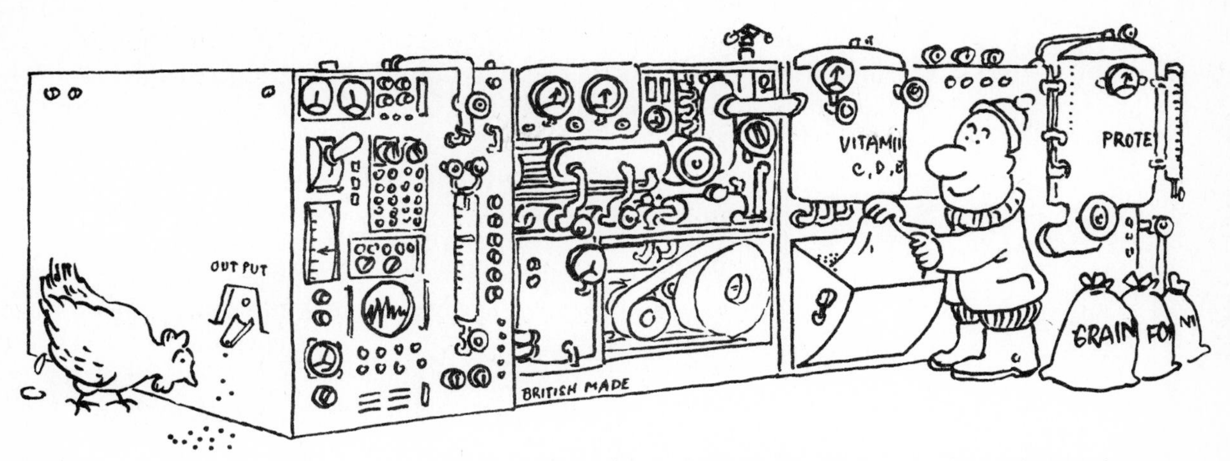

In a situation where two foods were being mixed – in quantities x and y, the region of 'allowable mixes' would usually be something like this.

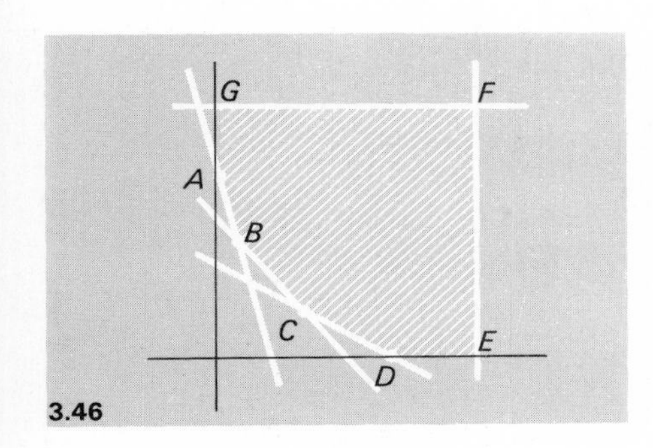

3.46

The constraints would be of the 'at least' type, so we would have to be 'above' the sloping lines. The lines EF and GF would correspond to some such constraint as the maximum amount of each type of foodstuff available or the quantity beyond which the chickens get a heart attack!

This time the optimum mix of feedstuff will be the one which achieves the required nutritional value and *minimizes* cost. Such a problem can be tackled by just the same method as before and the optimum arrangement will again correspond to one of the vertices. Just as a change in profit on each item could be easily accommodated in the previous case, this time adjustments for a change in the costs of feedstuffs can be made very simply. And this is quite a serious consideration. The prices of foodstuff fluctuate considerably and with this method, once the vertices are known the cost can quickly be calculated at each vertex whenever there is a change. This would be a perfectly routine operation once the problem had been set up.

We have spent some time on this type of problem for three reasons. It illustrates the advantage that coordinate geometry can give by providing a marriage between geometric notions and arithmetical problems. It has triggered off a few ideas that will be useful to us later. And it is an example of a technique used extensively in industry to solve problems of delivery, arrangements of resources, fertilizing farmland, and many other situations. It is an example of the *linear-programming* method, which came into its own in the 1950s and 1960s with the rise of electronic computers. Our example involved two variables and our geometric representation involved lines and regions of a plane. If we had been involved with three variables, we would have used planes instead of lines and regions of three dimensional space instead of regions of a plane. With more than three variables, the geometric notions need to be extended to more than three dimensions! This may seem rather daunting, but mathematicians are quite fond of fiddling about with geometry in n dimensions where n is unspecified, let alone greater than 3! Oil companies have very sophisticated computer programs for solving linear-programming problems with several hundred variables and constraints.

We shall not be going into the details of linear-programming techniques – it is a specialized subject that needs its own individual treatment. But the ideas that we used – specification of subsets of points by use of algebraic conditions, the notion of points belonging to more than one subset simultaneously, the use of different branches of mathematics as models of physical or mathematical situations, and the absolutely basic concepts of set and function – these are common to much of mathematics. In the next chapter we shall develop a little more formally some of these ideas that we have already used in an intuitive way.

4 Sets and functions

In this chapter we shall develop the two ideas of 'set' and 'function' that we used in the previous chapter. The words set and function are used in mathematics in very much the same way as in ordinary speech – the word 'set' means simply a collection of objects and the word 'function' is used in the sense of a dependence of one thing upon another – for example the rate at which central heating consumes fuel is a function of the temperature at which the house is to be maintained.

The two concepts have played key roles in the development of mathematics in this century. With hindsight one can see that this is hardly surprising. If mathematics is concerned to describe relationships between things, then the notion of one thing depending on another is likely to be a central one in the subject. And it also makes sense that it will be important to specify the collection of objects – or sets – between which the relationship exists. The natural question to ask, and the one asked by many who query the introduction of these topics into the school curricula, is 'how was it that so much success was achieved in mathematics before the introduction of these ideas?'

The answer to this is really in two parts. Mathematicians have always used the ideas, but only implicitly and in special cases.

We shall see in the next chapter that without an explicit use of the ideas it is difficult to explain certain features and deficiencies of the number system – difficulties that we hinted at in Chapters 1 and 2. Furthermore, until sets and functions were given lives of their own, their (implicit) use was restricted to relationships between the special cases of sets of numbers. Up to the later part of the nineteenth century, the only sets and functions used were essentially of the type we saw in Chapter 3, though perhaps more complicated.

In Chapter 3 we saw functions as relations between sets of numbers, specified by a formula, such as

$$y = x^2 + 3x + 2,$$

where each value of x specifies a value of y. But to say that a function is *specified* by such a formula is not to say what the function actually *is* – it simply describes what it *does*. Is the function the formula $y = x^2 + 3x + 2$, or the expression $x^2 + 3x + 2$, or what? In the past, the problem was dodged – nobody actually said what the function is; the type of phrase commonly used was 'if there is a formula connecting variables x and y such that, given values of x, corresponding values of y can be calculated, then we say that y is a function of x'. The drawback of this description is that it does not reveal a function as an entity in itself. It's like defining a bicycle pump as a device for blowing up bicycle tyres – it doesn't help you to know one when you see one. If you do not see this as a tremendous disadvantage, you should not be too distressed because, as we have said, it was not until late in the last century that the advantage was seen of regarding functions as objects, much like numbers, that could be manipulated – added, subtracted, and so on – as one might juggle algebraic symbols.

It is rather difficult to appreciate the important advantages that can be gained by this more general view of functions without looking at some mathematics that is more advanced than we are able to go into here, although you will see how the idea itself helps us to obtain an insight into the way mathematics works. So why have we bothered to bring it up? Because the modern definition of function is given in terms of sets and we offer it as reinforcement of the importance of sets, as an example of the efforts being made to base mathematics on as few concepts as possible, and we shall see that it gives a useful, dynamic, way of thinking of functions.

The second great advantage in an explicit use of the terms 'set' and 'function' is that it enables the ideas to be used in far more general contexts than numbers. It enables mathematical techniques to be employed in situations which are not just numerical – to describe and analyse the symmetries of a crystal, the logic of an argument, the structure of a network, or a genetic theory of evolution.

Sets

The *theory* of sets, as a subject in itself, is far from easy and it involves complicated logical and philosophical problems, especially when it is used as a basis not only for describing the nature of numbers but for developing acceptable methods of mathematical reasoning. We shall not be going into these; our concern will be with the more down-to-earth aspects of sets. It is important to realize that in most places where the word 'set' arises, even in university courses, the use is not much more than a notational device. It is a very powerful device, but in no sense does the use involve an intimate knowledge of set *theory*. Really all that is required is an understanding of what a set is and how different sets can be combined. If that does not seem very much, you are right – it isn't. But, as we shall later in this chapter, it does give a very powerful way of looking at some diverse topics, from the logical construction of statements, to the combination of electrical switching circuits.

So far we have talked only of sets of numbers, or sets of pairs of numbers. But the word set embraces more general things than that. A set is just a collection of objects. All that is required for a proper definition of a set is that, given any object at random, you can decide whether or not it belongs to the set in question. For example, the set of all Members of Parliament on 1 January 1893 is a perfectly well-defined set because given the name of any person alive on that day it should be possible to say whether or not he were a Member of Parliament on that day.

Equality of sets

How do you decide whether two sets are equal? You look at the objects in each set, and if the sets contain the same objects the sets are equal. So the set of 'all the days in the week' and the set 'Monday, Tuesday, ..., Sunday' are equal. So we're talking about equal sets here as different ways of expressing the same thing. It's as simple as that. But there is an interesting point to clear up concerning equality. Sets *in themselves* have no structure, no syntax – they are just collections and the order in which the collection is made is immaterial. For example the set of words in the sentence 'Dog bites postman' is the same as (equal to) the set of words in the sentence 'Postman bites dog'. The concept of a set is a primitive notion, just as a collection of words is a more primitive notion than a sentence. It is a significant mathematical activity to impose structure on sets and so build complicated structures from basic simple concepts.

Operations on sets

The collection of all the words on this page is a perfectly well-defined set. And so is the collection of words on the next page. Call the sets of words on this page A and the set of words on the next page B. Now these two sets are not the same, some words appear in one set but not the other, and these two sets can be combined to form yet more sets. For example the set of all those words that appear on *both* pages is a set formed from the previous two. And so is the set of words that appear on either or both pages.

Intersection is a 'refining' process. To belong to an intersection, objects must have the qualification for membership of *both* sets.

These two ways of combining sets are the ones that are used more frequently than others and they are given special names.

Given any two sets A and B the set of all objects that are in A *and* in B is called the intersection of A and B.

The set of all objects that are in A or in B or in both is called the *union* of A and B.

Union is an 'amalgamation' process–as long as an object qualifies for belonging to *either* set it can get into the union of the sets.

It is useful to represent these ideas pictorially. Pictures can save many words and are useful for the immediacy with which they convey ideas. They are used extensively in situations where sets are being used. For example, if we represent the objects of two sets A and B as points within closed boundaries, like this,

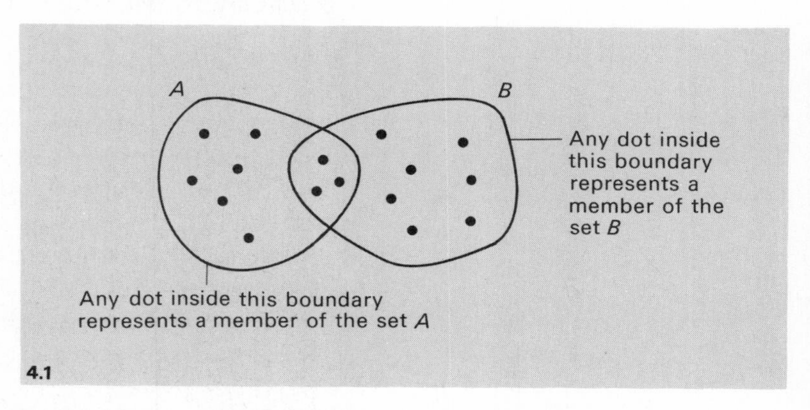

4.1

then it is easy to see what boundary encloses the set of objects that are in both A and B.

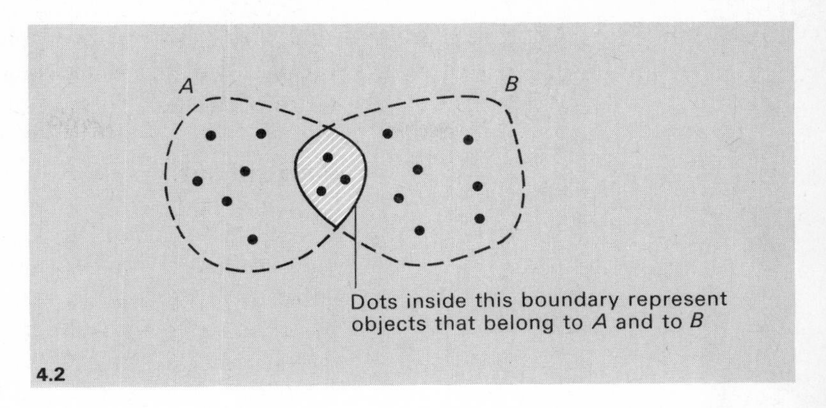

4.2

Points representing objects in the intersection of A and B will lie in the shaded region.

The region containing objects in the *union* of A and B can be represented in a similar way.

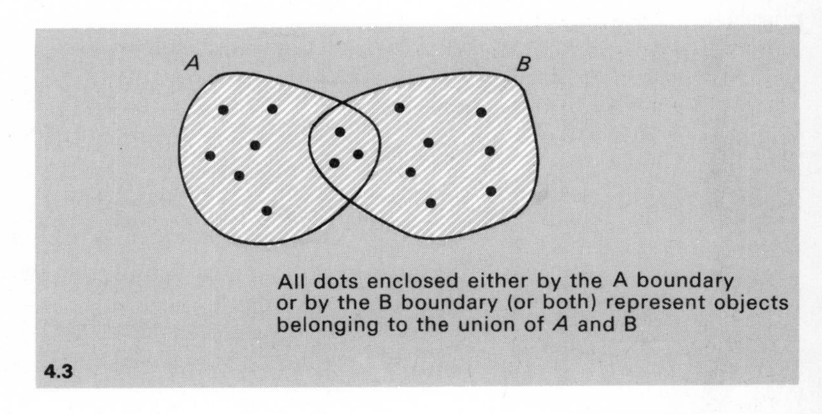

4.3

What sets do the shaded regions in the next diagram represent?

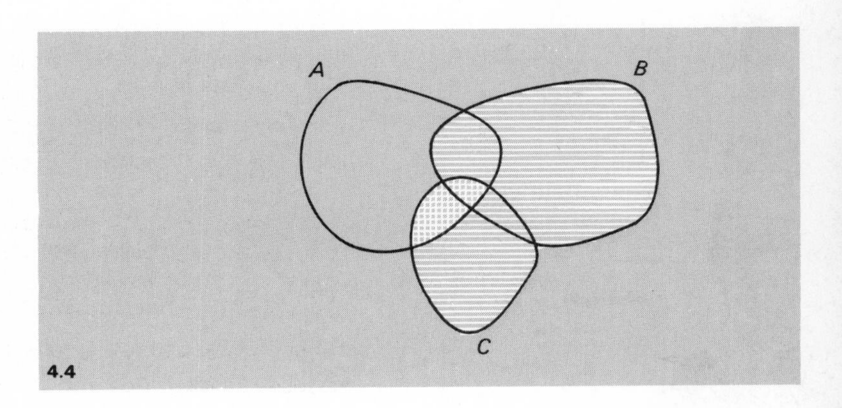

4.4

The horizontally shaded region represents the union of B and C. The vertically shaded region represents the intersection of A and C. (The vertically shaded region also has horizontal shading because to belong to the intersection of A and C an object *must* belong to C and this will automatically qualify it for membership of the union of B and C.)

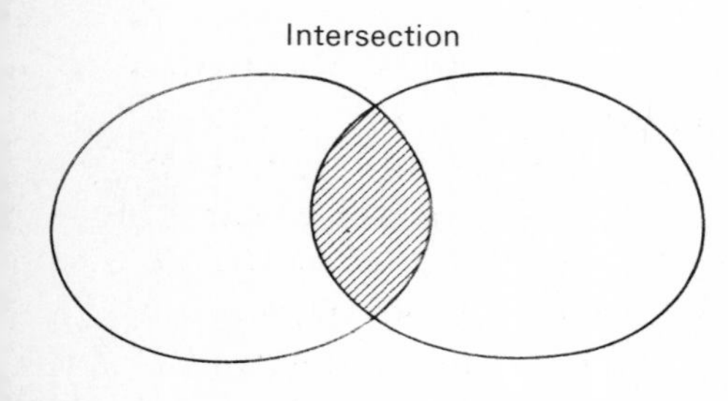

You do not have to use these diagrams. You can just use words. For example, if A is the set of all integers greater than 10 and B is the set of all integers less than 30, then the intersection of A and B is the set of integers between 10 and 30 and the union of A and B is the set of all integers: try representing this on a diagram if you like.

As a second example, let A be the set of names {Jack, Jill, Tom, Jerry} and B be the set of names {Mary, Tom, Robert, Elizabeth}. (The curly brackets that we have used to enclose the list of objects in these sets are the most commonly used symbols for this purpose.) Then the intersection of A and B is {Tom} and the union of A and B is {Jack, Jill, Tom, Jerry, Mary, Robert, Elizabeth}.

As a final example, if the set A is the set of all parents who have children at school, and the set B is the set of all parents who have children not at school, then the intersection of A and B is the set of all parents who have one child or more at school and one child or more not at school. The union of A and B is the set of all parents.

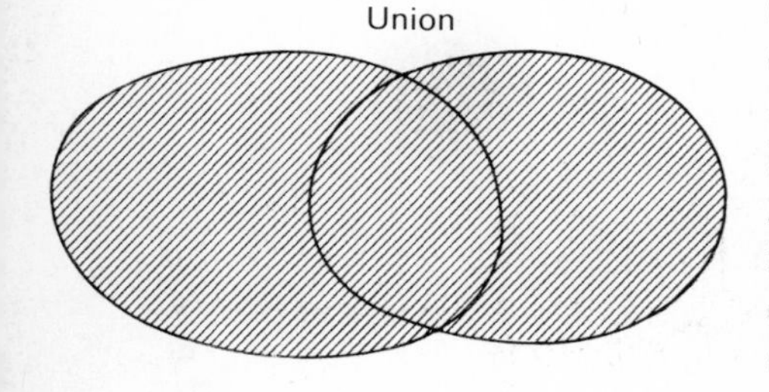

Obviously one could go on for ages making up simple examples like this. But the purpose of a simple example is to get the feel for a new concept before going on and using it in a more serious context. But even the simplest of examples can reveal important points. Suppose we take A as the set {Fred, Jane, Joan} and B as the set {Elizabeth, Mary, Robert}. The union of A and B (amalgamate them) is {Fred, Jane, Joan, Elizabeth, Mary, Robert}, but what about the intersection (what is common to both)? There is nothing common to both, so how do we say this in terms of sets? We say that the intersection is the empty set, the set containing no objects!

This is not a sophisticated notion that you need to worry about, it is just a device to keep things neat and tidy. If the operations of union and intersection are to be 'proper' ones, then we want them to give us an answer whatever pair of sets we apply them to – so we invent the term 'empty set'. The

Empty Sets are sets containing no elements. The intersection of the sets *A* and *B* below is the empty set.

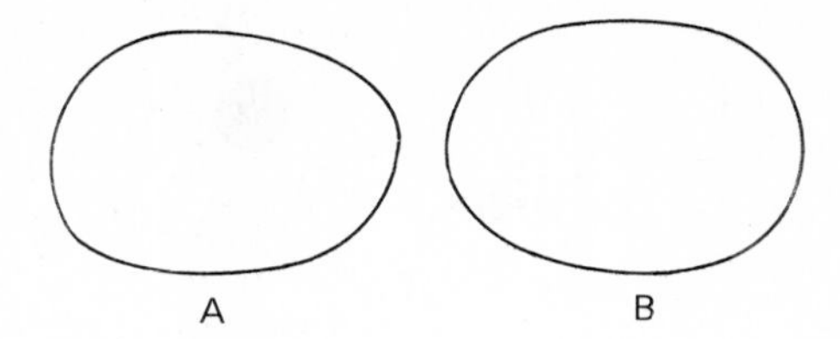

situation is rather similar to our preferring to say that $6-6$ is the *number* zero rather than 'nothing'. And just as you can do arithmetic with zero, so you can perform set operations with the empty set. You may like to ponder for a moment: can you see that whatever set A you choose, the union of A with the empty set is A and the intersection of A with the empty set is the empty set? Do you find it interesting to compare that with the facts that whatever number x you choose, $x+0 = x$ and $x\times 0 = 0$?

Notation

We shall be needing to use special symbols to stand for the operations of union and intersection of sets. In the first place, notation serves as a simple shorthand; in the way that we write + instead of the word 'add' for example. But once you have become familiar with it a good notation is more than an aid to quicker writing; it frees the mind to think about other things, for it should convey quickly several, maybe complicated, things all in one 'package'. Just think of the sophistication in the expression

$$\frac{3+7\times 6}{5}$$

which means $\frac{3+42}{5}$, which is to say $\frac{45}{5}$, which is just 9. The symbols not only shorten the writing; they also help to convey the meaning more immediately. Try conveying this expression over the telephone (just saying 9 won't do!); it is not exactly difficult, but is certainly prone to error (and you could easily think of something *really* difficult).

It has been suggested by C. H. Waddinton (in *Tools for Thought*) that the brain is capable of handling only about eight different concepts at any one time. This gives some indication of the extreme importance of being able to package together lots of constituent concepts into one 'higher order' concept that can be regarded as an entity.

$A \cap B$ means the intersection of A and B.
$A \cup B$ means the union of A and B.

For the operations of intersection and union we use rather strange symbols; we use $\cap$ for intersection and $\cup$ for union. So $A \cap B$ is a new set – the intersection of A and B, and the set $A \cup B$ is also a new set – the union of A and B. The usual way to read these expressions when you come across them in print is to read $A \cap B$ as 'A intersection B' and $A \cup B$ as 'A union B'. So $\cup$ and $\cap$ are *operations* which, when performed on two sets produce a third – in much the same way as + and × are operations which, when performed on two number produce new ones. Notice that + and × are operations, but $2+3$ and 4×6, for example, are numbers.

There is a lot that we could go on to say about these two set operations and there are other useful operations that could be introduced. But we are already in a position to apply the ideas to various situations, so let's do so.

Sets and logic

A number of mathematicians in the nineteenth century were interested in the problem of formalizing logic by using mathematical notation. The Greeks had laid down certain principles of logic and argument. Notable amongst these was Aristotle, a pupil of Socrates and tutor to Alexander the Great. He investigated the nature of politics and of biology and his principles of logic are still taught today. Of the people who started to develop his ideas of logic in the nineteenth century, perhaps the most famous are the Englishmen Charles Babbage and Charles Dodgson. Babbage was the inventor of the first calculating machine that could be called a computer in the modern sense of the word and Dodgson is famous amongst mathematicians for his mathematical work and amongst members of the general public for the books he published under his pseudonym of Lewis Carroll.

One of Dodgson's most famous problems goes like this:

> If all boiled red lobsters are dead, and all boiled dead lobsters are red, does it follow that all red, dead lobsters are boiled?

One way to tackle this is to consider three sets

R, the set of all red lobsters,

D, the set of all dead lobsters, and

B, the set of all boiled lobsters.

We can first translate the statement that all boiled red lobsters are dead. The set of boiled red lobsters is $B \cap R$ (boiled *and* red). If every lobster in this set is dead, then every member of the set $B \cap R$ must be in D. So on a set diagram, the set $B \cap R$ must be completely contained in the set D. So the diagram must look something like this – the shaded portion, representing boiled red lobsters, being contained completely in set D.

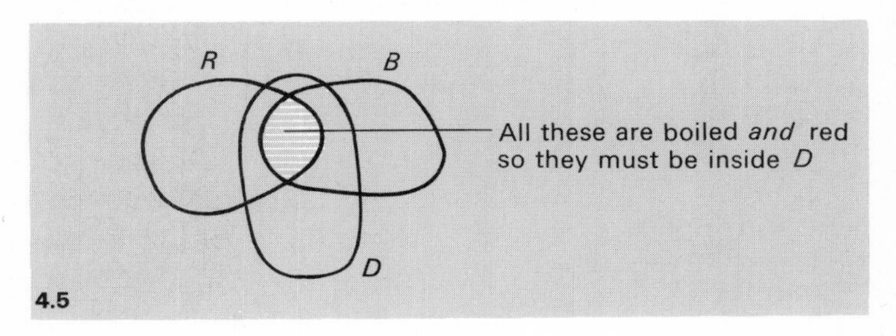

4.5

But we are also told that all boiled dead lobsters are red. This tells us that as well as everything in $B \cap R$ being in D, everything in $B \cap D$ must be in R. This can be realized in several ways. One possibility is like this:

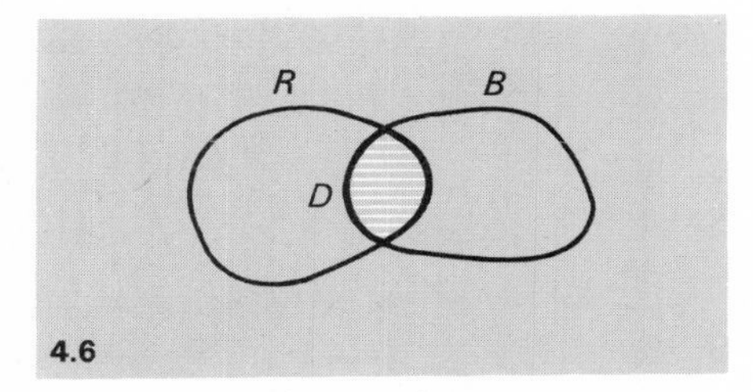

4.6

This way with $D = B \cap R$, gives the stated conclusion that all red dead lobsters are boiled.

But this arrangement

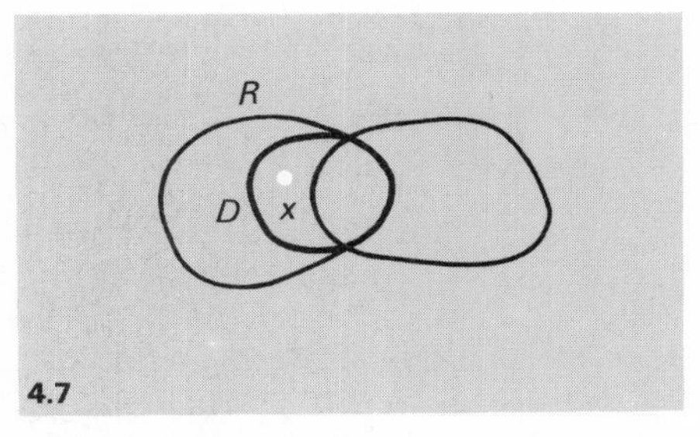

4.7

also satisfies the two conditions; and in this case the conclusion does not follow. The lobster x is red and dead, but not boiled.

The important point to notice is that by using ideas of sets we can actually use *diagrams* to help us to think through an essentially linguistic problem in an evocative and helpful way. It may seem at first meeting a rather strange way of thinking but there is no doubt that sketching diagrams – even in quite abstract situations – can be helpful.

You probably did not catch that argument on a first reading. If you are interested you might like to go back over it with a pencil and paper and follow it through for yourself. But even if you do not want to do this, can you see the possibilities? The effect of representing the problem in terms of sets, is to remove all the irrelevancies of the statement and make it easier to concentrate the mind on the essence of the argument. Of course, as is always the case using a new tool, it takes a little getting used to, but the job becomes much easier when the tool has been mastered. Even then, decisions have to be taken; is it really worth searching about in the shed for the extension lead for the electric drill or shall I just use a hand drill? Sometimes the advantages and disadvantages are pretty well-balanced; on other occasions a hand drill will just not do the job. My purpose here is just to show you that the tool exists, and to indicate some of the interesting points this raises. Before we proceed, however, it seems appropriate, since we are talking about logical argument, to say a few things about the idea of proof.

Proving things

It is important to realize first that a mathematical proof attempts to say nothing about ultimate truth. It tells you only that the conclusion is true if each of the hypotheses is true. For example, the following could well be a mathematical theorem.

Theorem: if London is in California then London is in America.

Proof: we know that California is in America, so if London is in California, London is in America.

This may seem rather silly, but there are two points worth noting. First of all the extra fact 'California is in America' is brought in to help the argument. This corresponds to the practice in mathematics of being allowed to assume any theorem that has previously been proved. You start off with a set of axioms, statements that are accepted as fundamental and not proveable, and from these you deduce statements. Any statement that can be shown to follow from the axioms by application of an acceptable set of rules of argument can be embodied into the collection of 'facts' and used in subsequent work with the same status as an axiom. Perhaps the most famous of all statements taken as axioms are those in the American Constitution: 'We hold it ....'. Euclidean geometry starts off with a set of basic assumptions (that are taken as unproveable), about points and lines, and then proves certain results from these axioms. Once it has been proved, for example, that the angles of a triangle add up to 180°, this result can be used in any subsequent argument. This begs the question, however, of what is meant by an 'acceptable argument' and the whole area of logic itself is a lively and active branch of mathematics.

The other point about our little theorem is to do with the validity of a result outside mathematics. However good is the internal consistency of a piece of mathematics, the conclusions are only as good as the hypotheses. If the axioms give a reasonably good representation of the things about us then the conclusions also should have reasonable interpretations. But that is about all you can say. Perhaps the clearest examples of this point arise in applications of mathematics in science. A certain hypothesis, say Newton's explanation of the laws of motion, is expressed in mathematical terms and then the processes of mathematics are used to deduce consequences of the hypothesis. If these consequences are reinterpreted back into physical terms, then they can be tested by experiment. The more consequences that are confirmed by experiment the greater is the confidence put into the hypothesis. But if experiments show that a consequence is false then, provided the mathematical chain of reasoning is correct, serious doubt is thrown on the hypothesis. Such a discovery can have a devastating effect on long, complicated chains of reasoning and demand a new hypothesis, as for example Einstein's theory – introduced to improve Newton's laws of motion.

Another point about proof is illustrated by the following theorem.

Theorem: some men are fat and some are happy, therefore some fat men are happy.

Each of these statements is certainly true, but the theorem itself is false because we are not told, nor can we *deduce*, that the set of fat men and the set of happy men have any men in common.

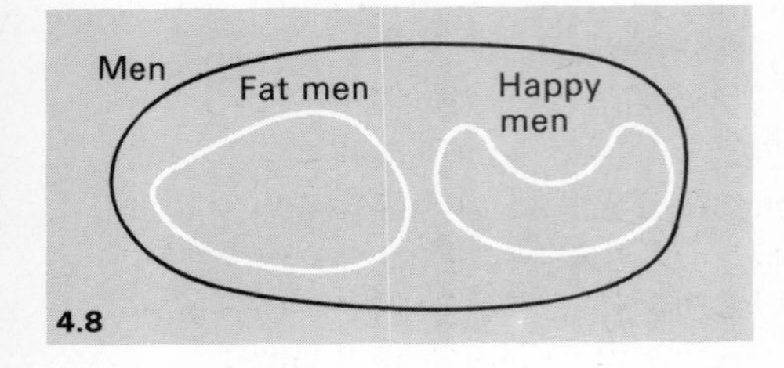

This diagram is entirely consistent with the first part of the theorem – it describes the statement that some men are fat and some are happy. It shows that this statement can be satisfied without the sets for fat men and happy men overlapping – in other words it shows that the statement can be satisfied without any fat men being happy. So even though each of the hypotheses is undoubtedly true and so is the conclusion, because each can be actually demonstrated, the theorem *itself* is not true.

Manipulating statements

We have seen that it can be helpful to use diagrams to help our thinking about both statements and sets. The fact that the type of diagram is the same in each case does, almost it seems by good fortune, seem to establish a link between sets and statements. We have also seen that we can combine two sets A and B, to form new sets, $A \cap B$ for example. The question we ask now is, given there is a link between sets and statements, is there a link between combining sets and combining statements? Take the statements

Albert is a teacher

Albert can do mathematics.

It is perfectly reasonable to combine these statements and to form a more complicated statement like

Albert is a teacher *and* Albert can do mathematics.

We can reinterpret these statements in terms of sets. If A is the set of teachers, then the statement 'Albert is a teacher' is equivalent to saying 'Albert belongs to the set A'. If B is the set of people who can do mathematics, then 'Albert belongs to the set $A \cap B$' is another way of expressing the compound statement 'Albert is a teacher *and* Albert can do mathematics'.

Symbols for statements

So there *is* a correspondence between combining sets and combining statements. What we ask next is, now that we have a way of representing statements mathematically, can we develop this idea a little further and actually manipulate statements in a mechanical, symbolic, way? You will see that we can and this will make a point about the effectiveness and power of mathematical notation and it will also reveal a surprising link between sets and how you might try to wire up a two-way light switch for a light on the landing!

One point before we get started. This section is not really in the mainstream of the development of this book, but it is included because it makes some important mathematical points of a general nature. It is these you should try to grasp, rather than attempt to gain an efficient technical facility. If you see the work that children are doing at school these days, you will also notice a familiarity with some of the topics that are offerred in many syllabuses.

The first step to developing a formal, algebraic approach to analyzing statements is to use symbols to stand for statements. For example, let's use the symbols a and b to stand for two statements that can be true or false (so a statement such as 'I wandered lonely as a cloud' is not the sort of thing we are working with). Then as the next step we need a symbol to combine these symbols a and b in a way that corresponds to forming the intersection of sets. The most common choice is $\wedge$. Then $a \wedge b$ represents the required composite statement. We usually read '$a \wedge b$' as 'a and b' – some people preserve the analogy with sets and read it as 'a intersection b'.

In everyday language we also combine statements using the conjunction 'or'. But the compound statement

Albert is a teacher or Albert can do mathematics

is ambiguous. Do we mean that 'Albert is a teacher or he can do maths but not both', or do we mean that 'Albert is a teacher or he can do maths or both'? Well, once the ambiguity has been spotted it ceases to become particularly important, for it can be settled just by agreeing to adopt a convention for always taking one of the alternatives. The most convenient convention is to take the 'inclusive' case – teach or do maths or both. This then corresponds to the union of two sets, where an object belongs to $A \cup B$ if it belongs to one or other or both of the sets A and B.

The symbol we use for this conjunction of statements is $\vee$, so $a \vee b$ (read 'a or b' or 'a union b') is the compound statement 'a or b'.

Establishing the connection

Let's spend just a couple of pages in really establishing the connection between sets and statements once and for all. The way we shall do it is to represent each of them in a new, but *common*, way. Suppose we have two sets, A and B, and an object. Four situations can arise. The object can belong to neither A nor B, it could belong to B but not A, it could belong to A but not B, or it could belong to both A and B. So for any object you pick, the four possibilities can be represented by a table. The technique is to use a 0 to represent the case when the object does not belong to a set and a 1 when it does. So we get this arrangement

	A	B
Object belongs to neither	0	0
Object belongs to B but not A	0	1
Object belongs to A but not B	1	0
Object belongs to A and B	1	1

Can we fill in a column for the set $A \cap B$?

A	B	$A \cap B$
0	0	?
0	1	?
1	0	?
1	1	?

When will the object belong to $A \cap B$? Only when it is in both A and B. So the column for $A \cap B$ looks like this

A	B	$A \cap B$	
0	0	0	Object does not belong to $A \cap B$
0	1	0	Object does not belong to $A \cap B$
1	0	0	Object does not belong to $A \cap B$
1	1	1	Object *does* belong to $A \cap B$.

The column for $A \cup B$ looks like this

A	B	$A \cup B$	
0	0	0	Object does not belong to $A \cup B$
0	1	1	Object does belong to $A \cup B$
1	0	1	Object does belong to $A \cup B$
1	1	1	Object does belong to $A \cup B$.

because the object will belong to $A \cup B$ if it belongs to A or to B or to both A and B. If you can get that idea, there is not much more to it, all the rest is a matter of using this idea rather than introducing new ones. First let's see how the idea can be used for demonstrating equality of different combinations of sets. Consider, for example, the set $A \cup (A \cap B)$. We can construct the table like this:

A	B	$A \cap B$	$A \cup (A \cap B)$
0	0	0	0
0	1	0	0
1	0	0	1
1	1	1	1

The $A \cup (A \cap B)$ column is obtained by looking at the A column and the $A \cap B$ column and forming the union – again the object will belong to the union only if it belongs to one or other of the two sets.

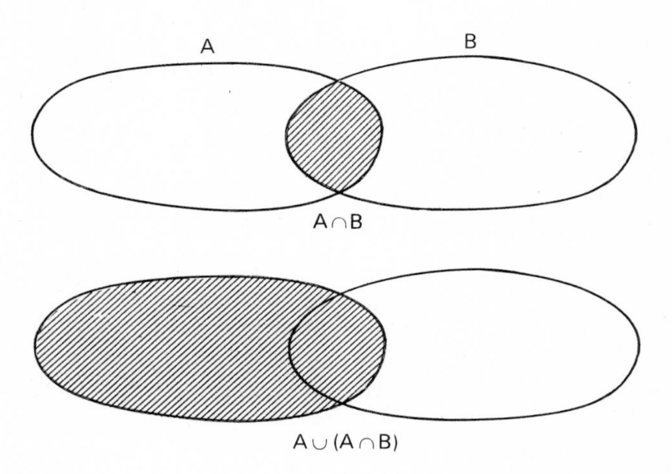

Now look at the $A \cup (A \cap B)$ column and the A column. They are the same, so whenever an object belongs to A it belongs to $A \cup (A \cap B)$ and whenever it belongs to $A \cup (A \cap B)$ it belongs to A. So the two sets are the same. So we have demonstrated – in a *purely mechanical way*, that $A \cup (A \cap B) = A$. You could have worked this out by a verbal argument, but for more complicated cases, a purely mechanical technique has considerable advantage.

Now let's apply the same idea to statements. Remember that the only statements we are considering are those which must either be true or false: just as an object does or does not belong to a set. All we have to do is go through the same procedure for statements as we just followed for sets.

Combining statements

Given two statements a and b, again four situations can arise – a and b both false, a false and b true, a true and b false, a and b both true; so again every possibility can be represented by a table:

a	b	
0	0	Both false
0	1	a false, b true
1	0	a true, b false
1	1	a true, b true

The statement $a \wedge b$, 'a and b', will be true only if both a and b are true, so it can be represented like this:

a	b	$a \wedge b$
0	0	0
0	1	0
1	0	0
1	1	1

$a \wedge b$ is true only if both a and b are true

And $a \vee b$ can be represented in a similar way

a	b	$a \vee b$	
0	0	0	a false, b false
0	1	1	a false, b true
1	0	1	a true, b false
1	1	1	a true, b true

$a \vee b$ is true if a is true or b is true or both are true.

A comparison of these tables with those for combining sets is a pretty convincing demonstration of the equivalence of the two situations.

Using these two methods of combining statements, we can form quite complicated statements. For example, if a, b, and c are statements, then

$$(a \wedge (b \wedge c) \vee c) \wedge b$$

is a statement with a very complicated logical structure. The advantage of representing such statements symbolically, is that once the rules for manipulating the operations are known, there is a chance that just by mechanical manipulation of the symbols it may be possible to reduce a complicated statement to a simpler form, just as by knowing the rules of arithmetic we can simplify the ordinary algebraic expression $a \times b + a \times c$ to $a \times (b+c)$.

To go into more detail would take us too far into a specialized subject but if you can see the general drift of what is going on you will be able to seize on an important mathematical principle. We are using symbols and what is apparently an abstract method of combining them to represent a concrete situation. The brevity and formality of the mathematical formulation enables us to concentrate far more closely on the essence, the structure, of the practical situation, and in that way put ourselves in a stronger position to analyse it. The mathematical system is a model of the concrete situation – a model that preserves only those aspects on which we wish to concentrate. In this way, mathematical notation is acting as more than just a shorthand: it is helping to communicate information by concentrating our minds. And by devising methods of combining the symbols, we are developing a language for manipulating and analysing the information.

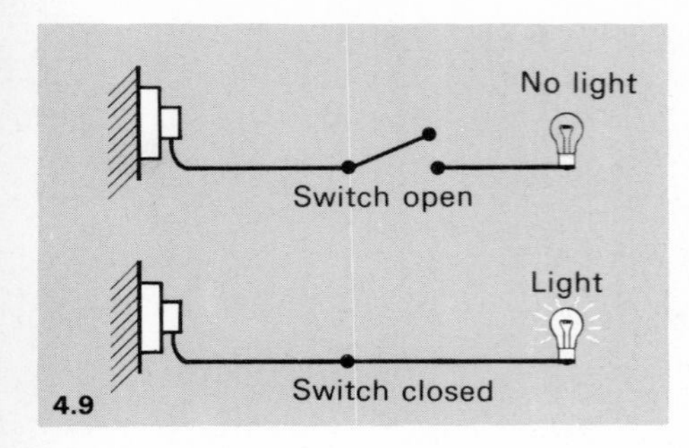

4.9

Switching circuits

You might well say that although there may be some intellectual satisfaction in seeing common features between sets and statements, it's all pretty abstract, and not really relevant to any practical situation. In fact, it is very useful to be able to model a logical process in this way. For example, computers rely on electronic circuits that exercise a certain logic. As a simple example of what I mean, imagine a wire along which an electric current can flow and which contains a switch that can be either open or closed. If the switch is closed, current will flow. If it is open, no current will flow.

You can connect two switches together and consider the various possibilities that can arise with both switches open, both closed, or one open and one closed.

Just as with sets and with statements the four possibilities can be tabulated. Let's call the switches p and q.

	p	q	
Both switches open	0	0	Current does not flow through p or q
p open, q closed	0	1	Current flows through q but not p
q open, p closed	1	0	Current flows through p but not q
Both closed	1	1	Current flows through p and q

Combining switches

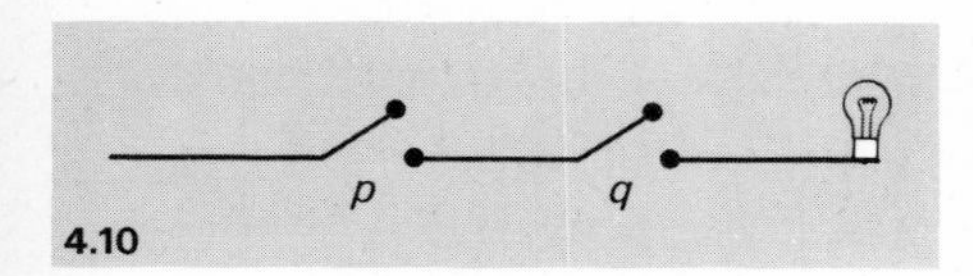

4.10

If you connect them 'end-on' like this then the light will go on only if *both* switches, p *and* q, are closed.

So you can regard the two switches as being combined together to form a single switch which is closed (passes current) only when *both* of the individual switches are closed. So the characteristics of the combined switch can also be represented by a table

p	q	Combined switch
0	0	0
0	1	0
1	0	0
1	1	1

4.11

If you connect them like this

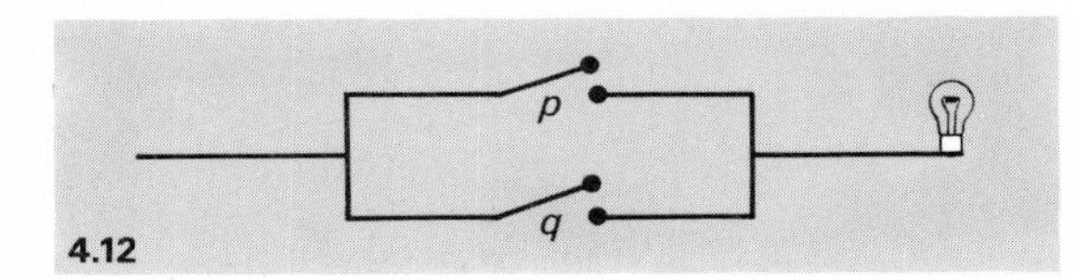

4.12

then electricity can flow along either branch, and the lamp will light if *either p or q* or both are closed. So the combined switch this time has the characteristics represented by the following table.

p	q	Combined switch
0	0	0
0	1	1
1	0	1
1	1	1

4.13

If you compare these tables with the ones we had for the operations on sets you'll see that the first, 'end-on' (in series) arrangement corresponds to the combination $\cap$ for sets or $\wedge$ for statements and the second arrangement (in parallel) corresponds to $\cup$ for sets or $\vee$ for statements. The relationship between the three systems is most appealing and, more than that, it illustrates a fascinating aspect of mathematics. The algebra of sets, the logic of statements, and the combination of switching circuits are all revealed as having the same essential structure and any problem posed in one of the settings can be transferred to one or the other settings. The mathematics acts not just as a model of a physical situation, important though that role is; it also acts as a link between physical situations so that experience in one area can be brought to bear on other, at first sight distinct areas. The mathematical model exposes the common ground and provides a channel of communication.

Suppose you have a rather complicated combination of the statements p and q such as

$$((p \vee q) \wedge p) \vee q.$$

You can build up a complicated switch to represent it. $p \vee q$ is represented just by

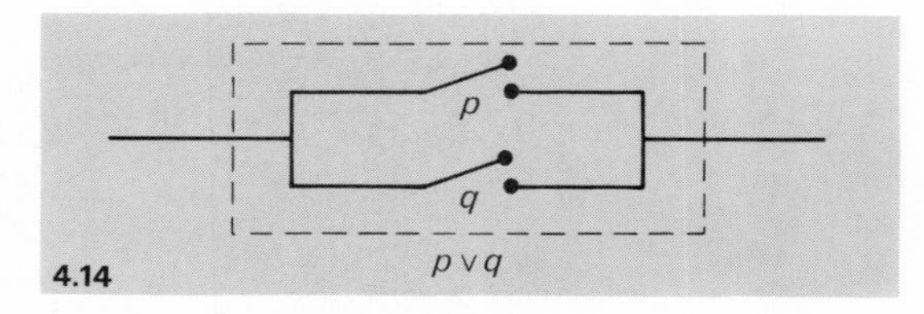

4.14

So $(p \vee q) \wedge p$ is the composite switch $(p \vee q)$ combined 'end-on' with p.

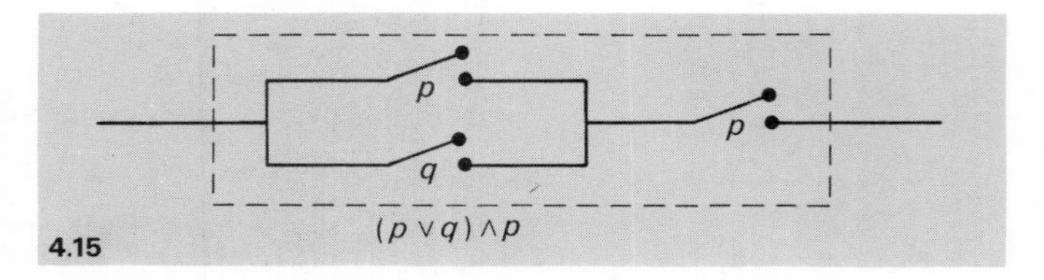

4.15

The fact that two switches are labelled p does not mean that they are physically the same switch, but that they are both closed or both open at the same time – in other words they are physically linked in some way. We now combine this complicated switch with q to form a switch to represent $(p \vee q) \wedge p) \vee q$.

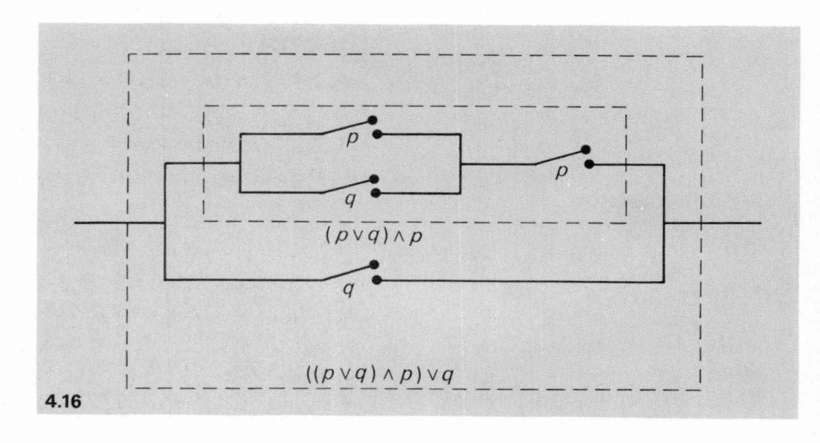

4.16

Look at the top branch of this circuit. Current can flow only if the right-hand p switch is closed. If this is closed, the other p switch must be closed and current can flow along the top branch of the $p \vee q$ switch. So if the p switches are closed, current will flow, irrespective of the state of the switch q. So the combined switch

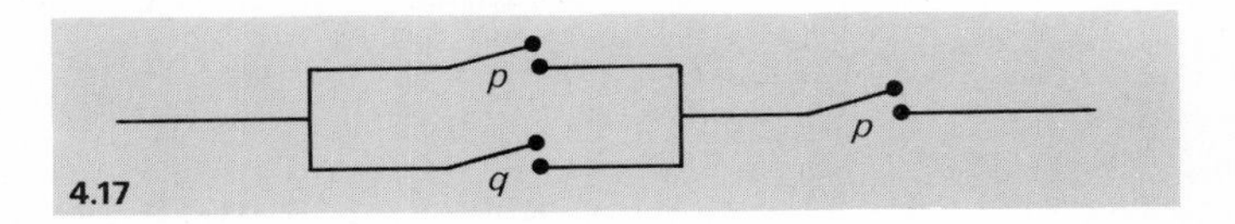

4.17

is equivalent to

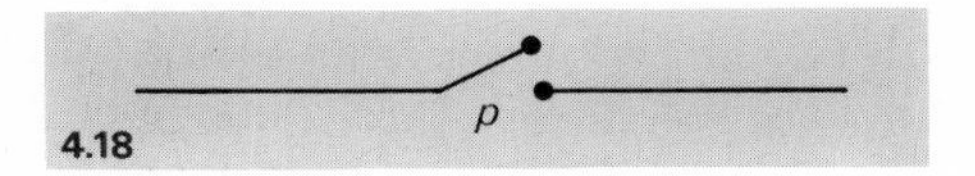

4.18

and the total switch is just

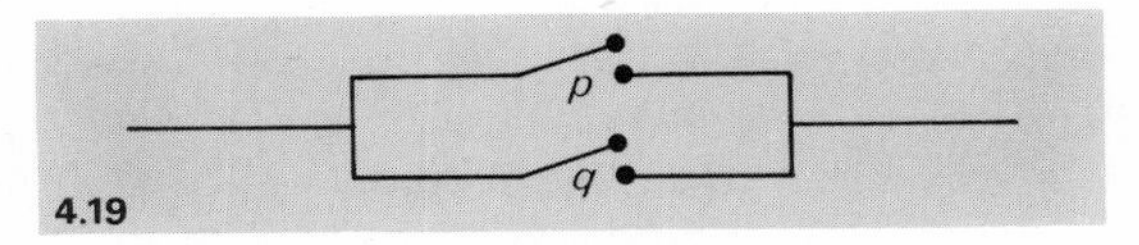

4.19

which represents the statement $p \vee q$. So the combined statement

$$((p \vee q) \wedge p) \vee q$$

is equivalent to the much simpler statement

$$p \vee q.$$

It is thus possible to simplify quite complicated statements by representing them as switches; a possible market here for helping people fill in their tax forms – or helping people design better ones!

But it is a two-way process. A great many machines today involve a lot of complicated switches and it is of interest to designers to try to simplify them. If you can represent one of these switches by an algebraic expression, then it may be possible to simplify that expression by straight-forward algebraic manipulation. And, talk about wheels within wheels, you can ask a computer, which is little more than a collection of complicated switches, to do the simplification for you!

Apart from the practical pay off, the ideas that underly these remarks are important in the way they characterize a way of mathematical thought. If several different situations have the same mathematical model, then this model reveals that they have essentially the same structure. Of course, a switch doesn't *look* like a statement, but if you are interested mainly in the way that they combine then that does not concern us. In fact if a model were a *completely* faithful representation of the entity being modelled, then there would be little point in making a model in the first place, for the model would be just as complicated and difficult to handle as the original.

If you want to set up a model, then the first thing to do is to decide what you are primarily concerned about and try to model that aspect faithfully – what happens to the rest is immaterial; preferably you want to be able to discard it, if that is possible. It is rather like the situation that confronts an artist. Different painters, asked to paint a tree might come up with all sorts of results. Some will try to reproduce as accurately as possible everything that the eye can see. Others will produce something far less representational, much more abstract. But they will all want to preserve what they regard as the essential structure of the subject, whatever they see as being the 'treeness' of the tree. It is the same with a mathematical model, whatever aspect is considered to be important to the particular investigation, it is that feature that must be preserved. The rest we would ideally like to recede into innocent inactivity.

Functions

We turn now to functions to try to fulfill our promise to say precisely what we mean by the term and to show how the idea can be expressed concretely in terms of sets.

Let us think for a moment about the function mentioned right at the beginning of the chapter, the one 'specified by the formula $y = x^2 + 3x + 2$'. If x takes value 1, then

$$y = 1^2 + 3 \times 1 + 2 = 6$$

If x takes the value 2 then y is 12, and so on. The formula specifies a set of *pairs* of numbers (0, 2), (1, 6), (2, 12), (3, 20), (1.3, 7.59) and so on. In any of these pairs, the first number is the value assigned to x in the formula and the second number is the corresponding value of y produced by the formula. We have already seen this idea in Chapter 3 in the context of drawing graphs: each of the pairs of numbers can be represented by a point. The set of all possible pairs of

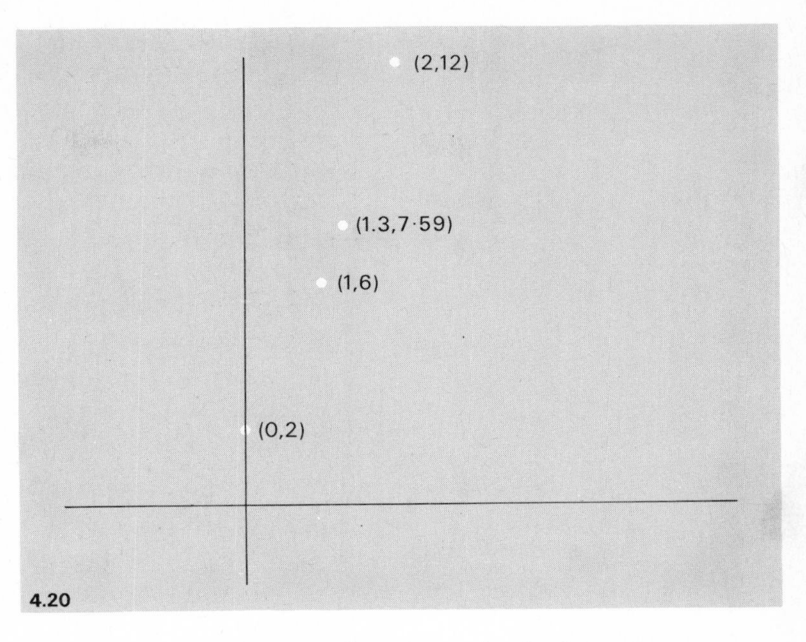

4.20

numbers, corresponding to letting x, in the formula $y = x^2 + 3x + 2$, take all the values it is allowed to take gives us a set of points – the graph of the function. Try drawing it yourself – after constructing a table of values of x and y. (Try using the values -3, -2, -1, 0, 1, 2 for x.) You should get something like this.

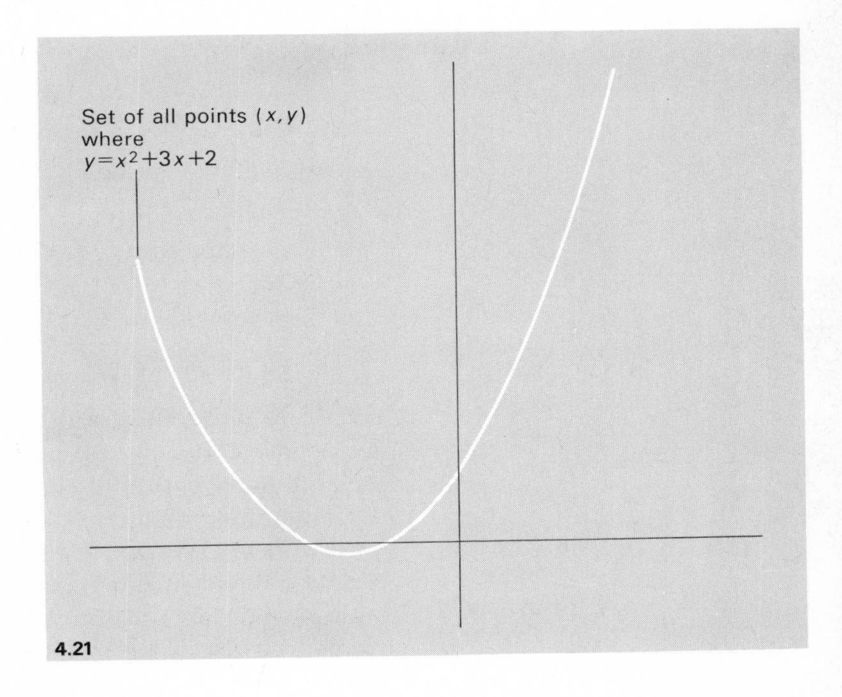

4.21

As we have said, we want to make the notion of function more concrete than a generic term for some vague kind of relationship between numbers. The graph goes a long way towards this goal, after all you can at least draw it. But the graph is just a visualization of the *set* which comprises these particular pairs of numbers, so in a sense this set is a more basic concept. Furthermore, we shall see examples of functions which relate sets other than sets of numbers – where the pairs are pairs of objects, not necessarily numbers. For such functions you cannot draw graphs, but the set of pairs of objects remains with us. It is this idea then, of a set of pairs, that we are going to take as our basic concept and it is with such sets that we are going to identify the term 'function'.

The crux of the matter is that the function completely specifies the set of pairs and the set of pairs completely specifies the function. It is a short, but significant step to establish the complete identification and *define* the function as the set of pairs. Let's have a down to earth example to get you used to the idea. Suppose a city planner is interested in the population of the capital cities of Europe. Essentially he is concerned with a function – the one that relates a number with each city. The function itself is the set of pairs

{(London, 8 180 000), (Paris, 2 811 000), ..., (Rome, 2 379 000)}

just as the set

$$\{(0, 2), (1, 6), (1.37, 7.59), \ldots\}$$

is the function specified by the formula

$$y = x^2 + 3x + 2.$$

The point is this; if either of the objects in the pair is not a number, you cannot have a formula. For example the objects in the pair (London, 8 180 000) cannot be connected by a *formula* only a verbal statement. Here is another example. When you look through a horticultural catalogue to choose a plant you have all sorts of things in mind – the flowering season, the size, the colour, and so on. Suppose for the moment that you are particularly interested in colour, then you will mentally pair each plant with its colour: (daffodil, yellow), (bluebell, blue), (rose pascali, white), and so on. The complete set of pairs is a function relating plants to colour. In a sense, this is a nonsense example – obviously it is not necessary to know anything about functions to plan your garden – but it does illustrate two important points. In the first place, it shows how commonplace the idea is and secondly how the definition of a function as a set of pairs is far more general than a specification in terms of a formula, such as $y = x^2 + 3x + 2$, or even in terms of graphs; for some functions, like our flower example, it is not possible to draw a graph. In later chapters you will meet more significant examples of this type; for the moment our purpose is just to get the ideas clear.

It is one thing to be able to define the general term 'function' but how do you actually specify a particular function? Suppose the pairs that make up the function are drawn from sets A and B. For example, in our population example, A would be the set of all European cities and B would be the set of all numbers. In the horticultural example, A would be the set of all flowers in the catalogue and B would be the set of all colours. In each case, if a is an object in A and b is an object in B, then the pair (a, b) is a candidate for inclusion in the set of pairs specifying the function. For example (daffodil, purple) is such a candidate, but it has to be excluded from the set of pairs constituting the function, because there is not a purple daffodil in the catalogue.

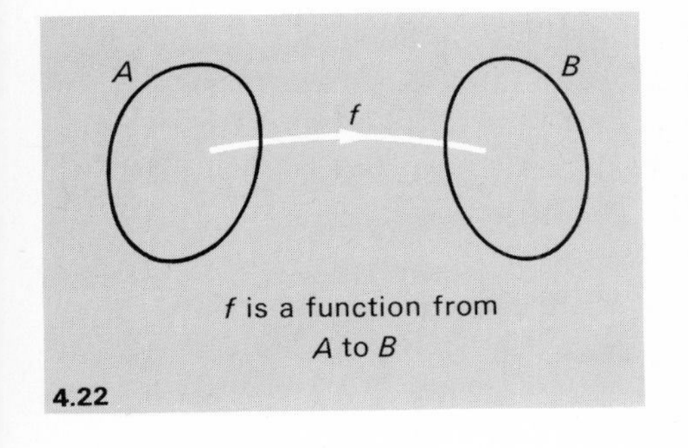

f is a function from A to B

4.22

Any function is a *subset* of the set of *all possible* pairs like (a, b) where a is drawn from the set A and b is drawn from the set B, just as a graph is a subset of all the points in the plane. It is sometimes useful to emphasize the notion of association of an object in B with an object in A and to do this we talk of f as being a function *from* A to B.

The problems of specifying a particular function is equivalent to specifying the subset of pairs. In the case of the flowers example, or the capital cities, you can write down the complete list of pairs. Alternatively, the rule by which the pairs are formed could be quoted. Thus we could refer to

> the function consisting of the pairs obtained by assigning to each capital city of Europe the number of people in its population.

This is the only method available to us when the sets involved contain an infinite number of elements, for then it is impossible to write down the complete list of pairs. The most important example of this type is when the sets are sets of numbers. In such cases it is often possible to express the condition which specifies the subset of pairs as a formula. Thus, just as we described graphs in the previous chapter, we may refer for example to the function

$$\{(x, y): y = 2x^2 + 1\}.$$

(Remember that you read this as 'the set of all pairs (x, y) such that $y = 2x^2 + 1$.) To be precise we should also specify the sets from which x and y are to be drawn. Failing to do this can sometimes lead to considerable difficulties, but in the cases we shall be considering, this extra information can quite safely be taken as implicit.

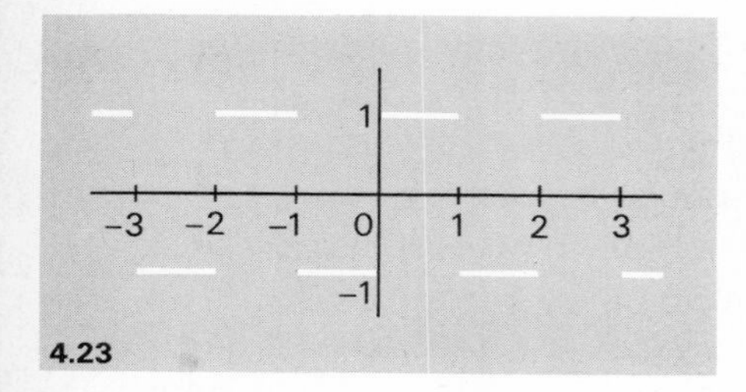

4.23

It is by no means essential to have a formula connecting the elements in the pairs. For example the function whose graph looks like Fig. 4.23 is important in some engineering contexts, but there is no simple formula relating the values of y with the values of x. So not all number pairs have a formula.

A dynamic view of functions

The definition of a function as a set of pairs has the advantage that it expresses functions in terms of the more fundamental notion of sets. This is an advantage from the point of view of the development of mathematics because it does not bring into consideration any essentially new concept. If the mathematical community can accept the notion of the set, then it can accept that of function, because it is just a special type of set. However, from the point of view of getting a real feel for what functions do for you in mathematics, it has certain drawbacks. Most important of these is that it does not bring out the *dynamic* aspect of the idea.

To see what is meant by this remark, look again at the function

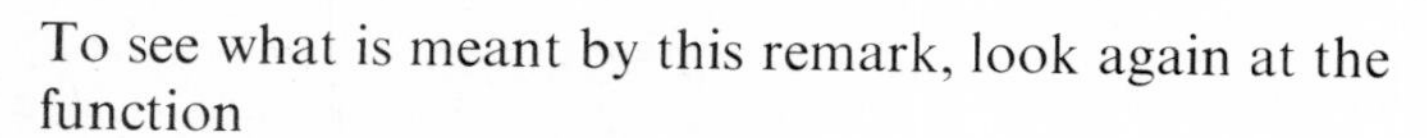

$$\{(x, y): y = x^2 + 3x + 2\}.$$

The graph of the function looks like this

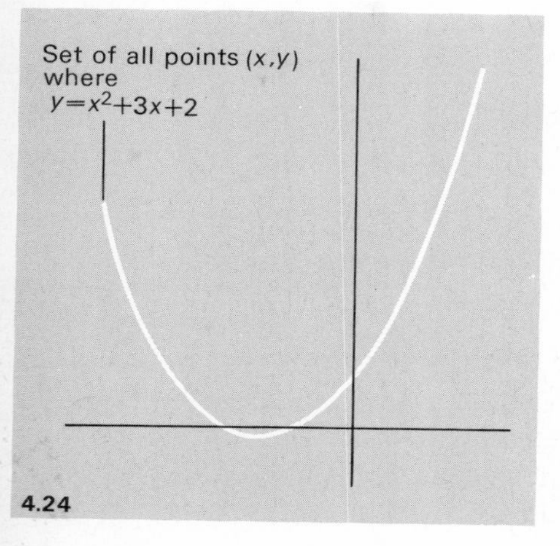

4.24

The function can be thought of as relating points on the x-axis with points on the y-axis. For example, the point 1 on the x-axis is related to the point 6 on the y-axis.

The arrows in the diagrams indicate this relationship in a dynamic way, and this is taken a step further by an evocative

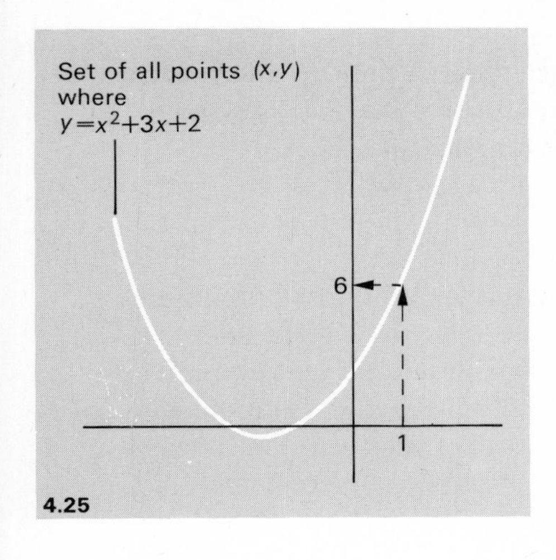

4.25

choice of words. We say that the function *maps* 1 to 6 and we write $1 \to 6$. This is just another way of saying that the pair (1, 6) is one of the pairs in the function, but it does evoke a useful way of thinking. If a function is going to be used extensively in any particular context then it is convenient to be able to refer to it. Some of the most common functions have special names, but for others we just use a symbol. For example, we might like to label the function above with the symbol f. Then, we could write

$$f: 1 \to 6, \text{ or } f: 3 \to 20, \text{ for example}$$

and read this just as 'f maps 1 to 6' or 'f maps 3 to 20'.

This form can also be used when specifying a function, by writing, for example

$$f: x \to x^2 + 3x + 2.$$

This tells us that any number, x, maps under f to the number $x^2 + 3x + 2$. This specifies the function because given any number, we can assign this value to x and calculate the number to which it maps. (As with the ordered pair method of specification, we should really mention explicitly the set from which the values of x are to be drawn – if we were to be absolutely precise we would want to distinguish between the cases when say x could be *any* number and when x were constrained to be positive, for example, and insist that these are different functions.)

This notation is important because not only is it rather easier than the formulation as a set of pairs, but it leads easily to an interpretation of function that indicates something of its practical importance. It indicates how a function can be thought of as a 'processor' – it takes one object and produces for us another one. The function f above takes the number 1 and gives us the number 6. The flower example we had earlier takes 'daffodil' and gives us 'yellow'. You can picture it like this if we label the 'flower function' g.

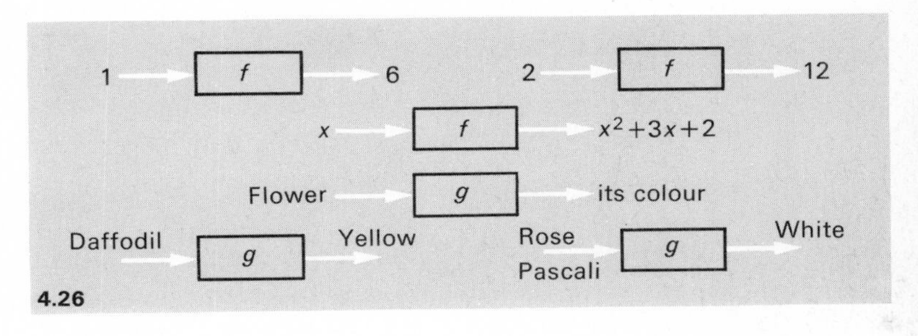

4.26

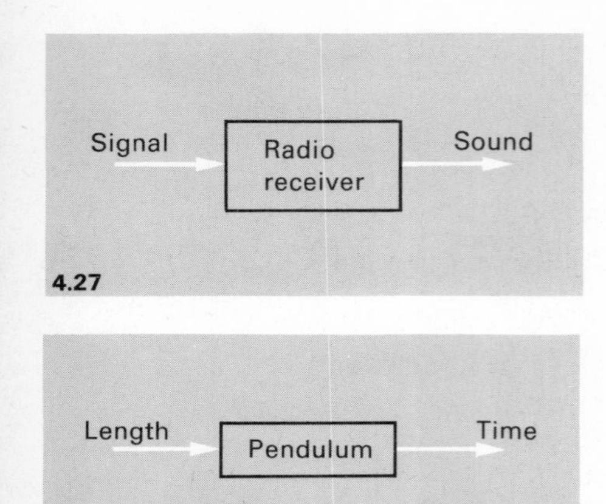

4.27

4.28

This is rather akin to the idea of a 'black box', that engineers find useful. A radio receiver, for example, may be a very complex piece of machinery but regarded as an entity it is a device that takes in an input – a radio signal – and produces an output – a sound.

The radio can be thought of as a function which associates particular sounds with particular signals. The idea is especially relevant in the experimental sciences. For example, suppose you were investigating the way a pendulum behaves. You might take a string with a weight attached and time the period of oscillation for various lengths of string. The pendulum can be thought of as a black box.

The object of the experiment is to try to find out how the black box works – to know the exact relationship between the input, length, and the output, time. In other words we want to find an exact specification of the function.

Any 'cause-and-effect' type of experiment is of a similar nature: if it is thought that a relationship exists between two sets of objects, then any search for a deeper understanding of that relationship is essentially a quest for the specification of a function.

Perhaps you are beginning to get an inkling of the ubiquity of the idea, but there are less obvious examples of functions. One of these we have already used implicitly. Do you remember this diagram on p. 57 in Chapter 3?

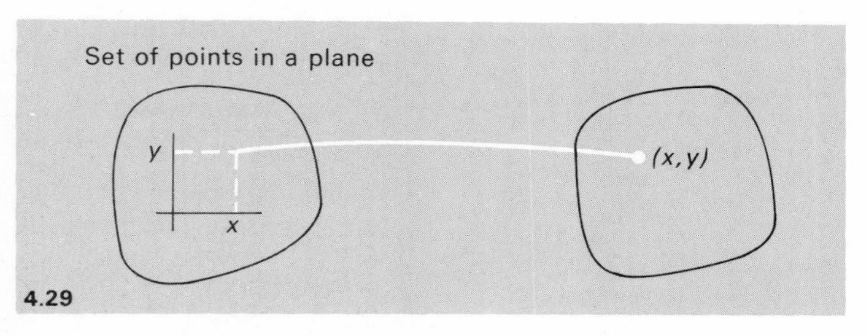

4.29

We used it when we were talking about using coordinate axes to establish a correspondence between points in the plane and pairs of numbers. The process is itself a function; one that maps points to pairs. Now don't worry about it too much, but give just a little attention to this thought: the subset of pairs $\{(x, y) : y = x^2 + 3x + 2\}$ maps to the subset of points consisting of those points lying on this curve

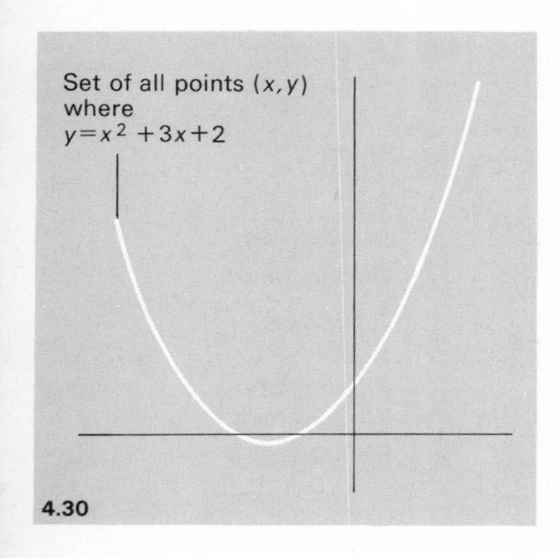

4.30

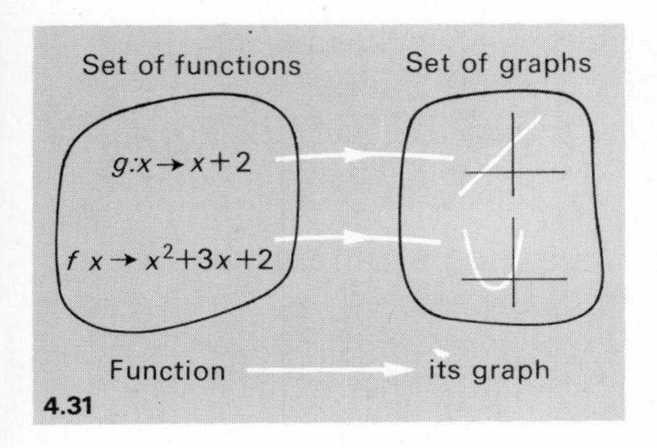

4.31

But the subset of points is itself a function, so we have a function that takes as its input a function – its output is a graph!

The details here are not important. What is important is to try to get an inkling of the potential of the idea of function.

Just two more examples, of rather different characters, may help.

Earlier in this chapter, we saw that the analysis of certain types of statement could be performed by algebraic manipulation, using letters to stand for statements and the symbols $\vee$ and $\wedge$ to stand for two ways of combining statements. The mathematical structure models the linguistic situation. We have two sets – a set of statements and a set of symbols (letters) and the mathematical model connects the two sets by assigning symbols to statements. So the mathematical model is another example of a function. Actually, it is a particularly powerful type of function because not only does it connect the statements with the symbols, it also connects the methods of combining statements with a way of combining the symbols.

Mapping shapes

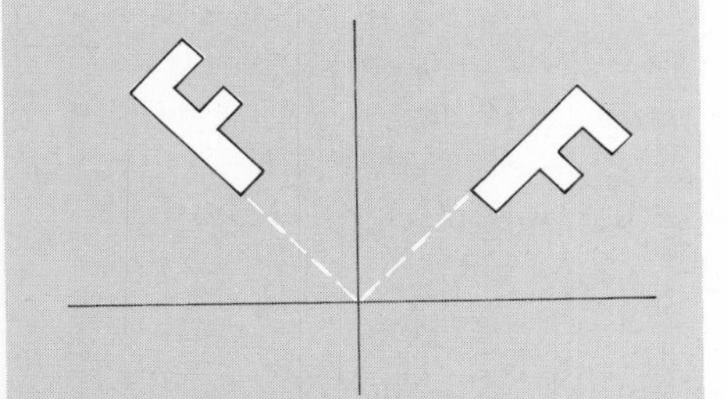
4.32

Our next example is again to do with points in the plane, but it is rather more geometric in nature than earlier ones.

The idea of 'transforming' points in the plane by some geometric process, like rotation or reflection for example is a useful device in many applications. We shall see a little of this later; for the moment, let's get clear what we mean by 'transforming'.

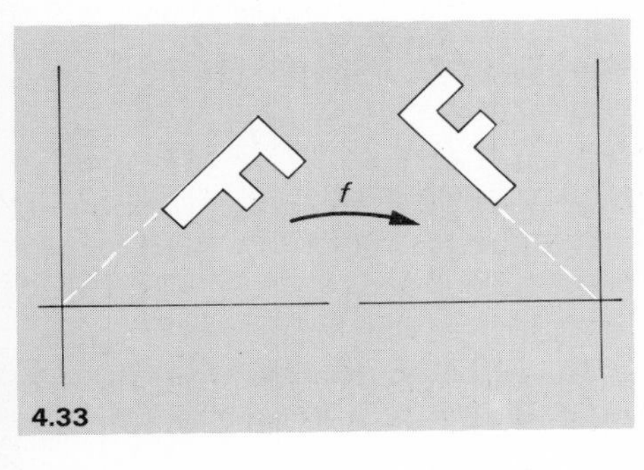

4.33

The relative positions of the two symbols in Fig. 4.32 can be described by saying that one can be 'mapped' to the other by a rotation through 90°. So if we think of a function that takes as its input the points in the plane and as its output these points all (except 0) with their positions changed by rotating through 90° anticlockwise, then the function maps the set of points making up the symbol to another set of points – also making up the symbol.

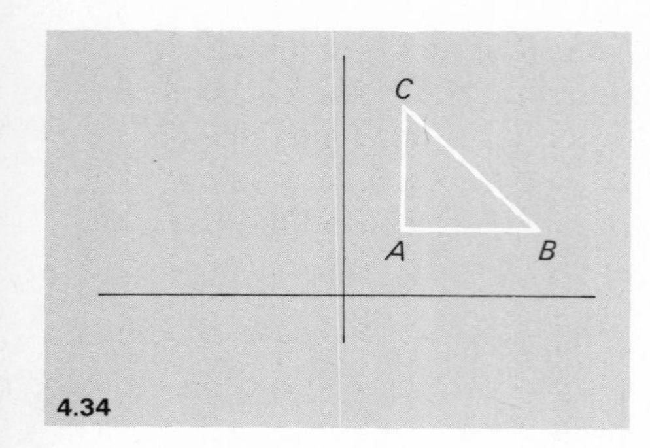

4.34

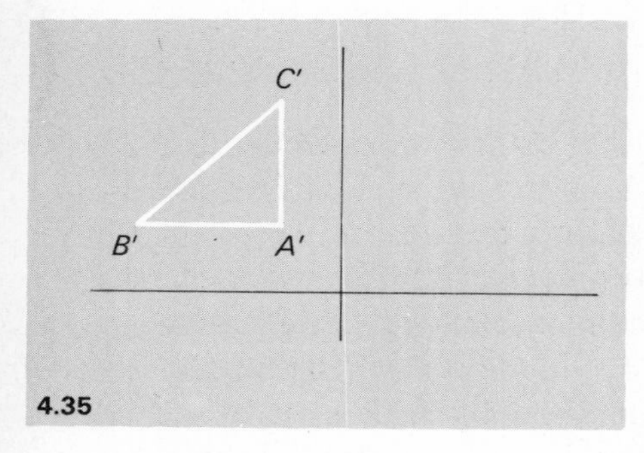

4.35

(You can actually describe this sort of function algebraically; it turns out in this particular case that if f maps the point (x, y) to the point (x', y') then

$$x' = -y \text{ and } y' = x.$$

We shall not be needing this sort of algebraic description but you may be interested to know that it exists.)

Functions like this are often called 'mappings of the plane to itself' – a description which although on close examination is rather difficult to comprehend does on a quick reading convey quite well the nature of the function, and that is the level of meaning which is intended. Another useful example of such functions is the one which reflects every point in the plane in a given line. For example, a reflection in the y axis, maps the triangle in Fig. 4.34 to the one in Fig. 4.35; A, B, and C being mapped to A', B', and C' respectively.

Describing symmetry

The type of function that maps shapes in the plane is useful when trying to give a mathematical description of symmetry. One of the concerns of mathematics is to describe the physical world and there can be no doubt that symmetry is one of the most remarkable of phenomena. Many shapes demonstrate symmetry and we all know what is meant by the term but, rather like an elephant, even though we know it when we see it it is rather more difficult to say exactly what it is. And then, what is the difference between the symmetry of, say, a square and a snowflake and an equilateral triangle?

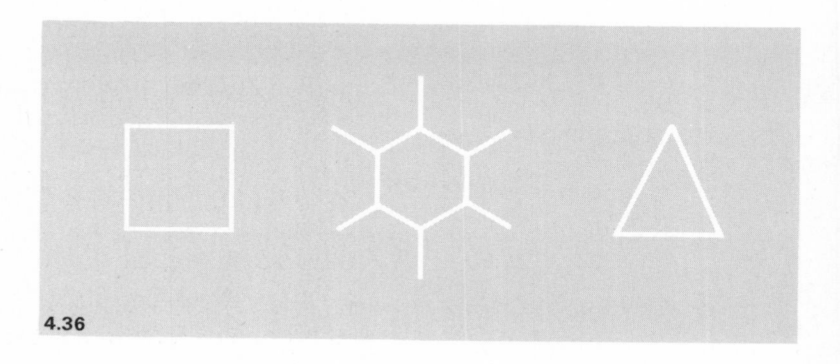
4.36

Functions enable us to do just that *and* they enable us to analyse the symmetries. Let's take a rectangle as an example.

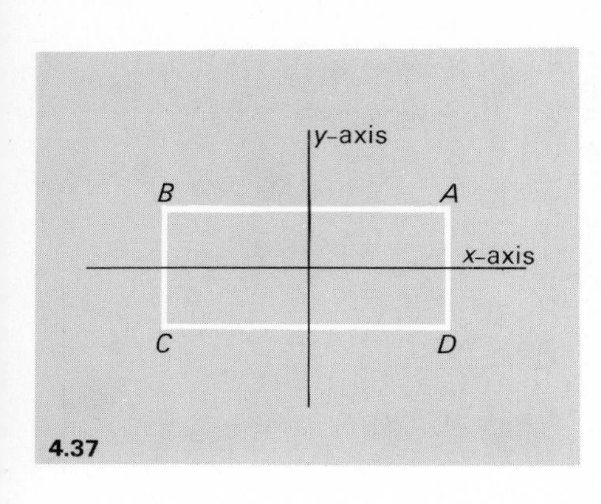

4.37

Imagine the corners A, B, C, and D as points in a plane, with axes as shown in Fig. 4.37. Then if all the points in the plane are reflected in the y-axis, the points A and B interchange and so do the points C and D. So if f is the function that maps each point of the plane to its reflection in the y-axis, we have

$$f: A \to B,$$

$$f: B \to A,$$

$$f: C \to D,$$

and $f: D \to C$.

Thus, although the points lying at the vertices of the rectangle are all moved by the action of the function, they are mapped to points which still lie at the vertices of the rectangle. Similarly, all the other points on the rectangle move, except those on the y-axis, but they still stay on the rectangle. In that sense, we can say that the effect of the function f is to map the rectangle to itself. Obviously, not every function that maps the points of the plane to themselves leaves the rectangle invariant in this way.

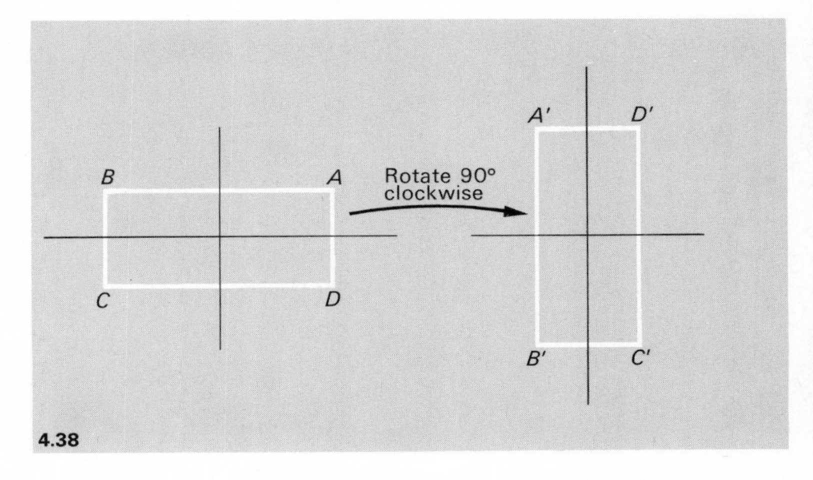

4.38

The fact that f does leave the rectangle invariant is a consequence of the symmetry of the rectangle – as indeed is the fact that the rotation through 90° does not leave it unchanged. There are other functions with the same property, for example the function that reflects all the points in the x-axis. We can describe the symmetry of the rectangle by listing all the functions that have this property – all the functions that leave the rectangle invariant. There are three functions like this.

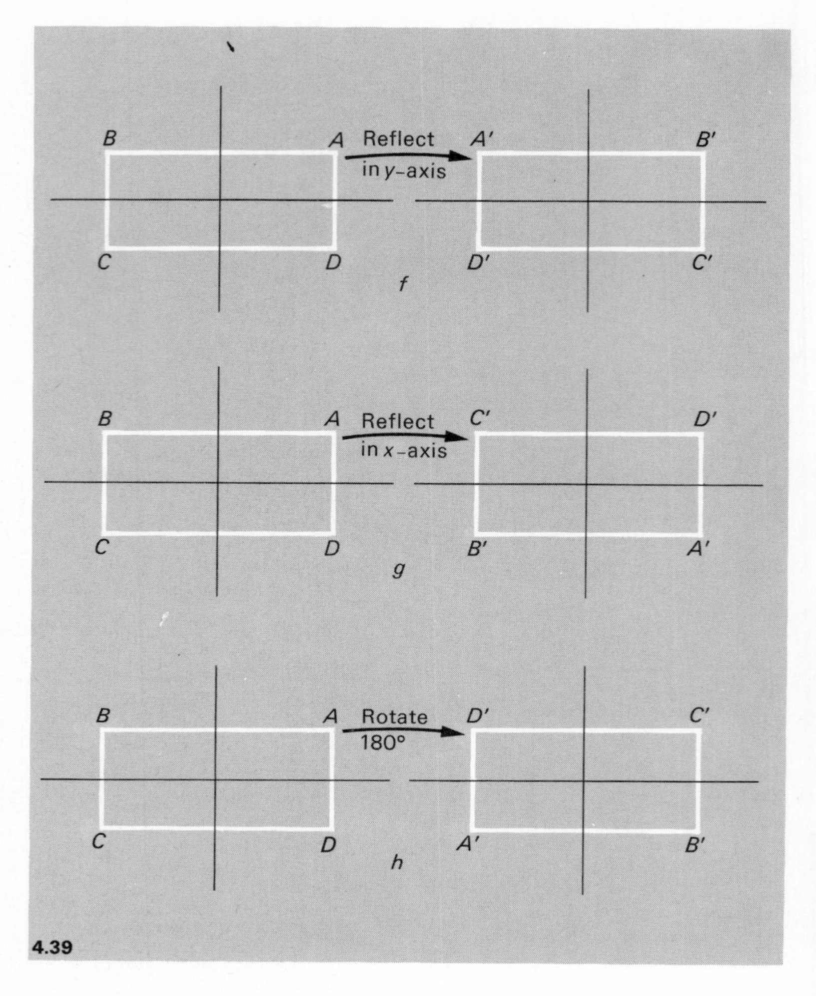

4.39

and one rather special one – the 'do nothing' function that leaves every point where it is.

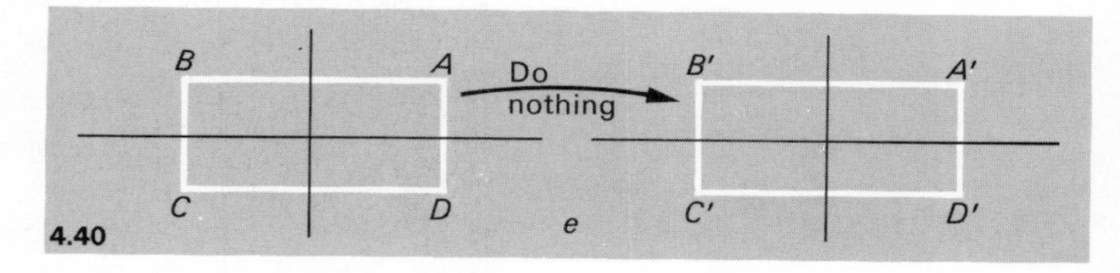

4.40

This may seem a rather trivial function but we shall see later why it is important. For ease of reference we call these four functions f, g, h, and e the symmetries of the rectangle and the set of four functions, $\{e, f, g, h\}$, describes the symmetry of the rectangle. Notice that the function e is a 'symmetry' of every figure. If the set of all symmetries of a figure is just $\{e\}$ then the figure is not symmetric!

What about the square? How do we describe *its* symmetry? All the functions that leave the rectangle invariant will also leave the square invariant. But so will the functions that rotate the points of the plane through 90° and through 270°.

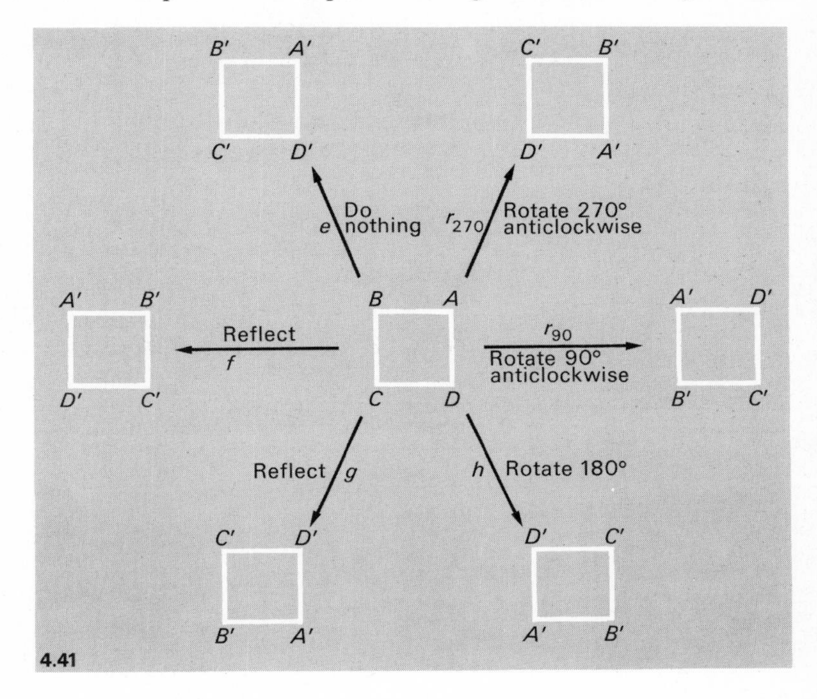

4.41

So the symmetry of the square is described by the set of six functions $\{e, f, g, h, r_{90}, r_{270}\}$. The difference between the symmetry of the square and the symmetry of the rectangle is described by the difference between the two sets of functions. In much the same way we can describe and distinguish between the symmetries of an equilateral triangle, or a snowflake, or a crystal, or whatever. If your mathematical instinct is working really well, you will begin to see that the degree of symmetry is connected to the number of functions – the 'more symmetric' the figure the more functions you would have. (And a circle would have an infinite number of functions – in a sense it is the most symmetric of all figures.) Later in this chapter, we shall take up these examples again and show how we can go even further towards distinguishing between the symmetries of geometric figures. But before we can do that we need a new idea – one that is both important and illustrates the remarks we made earlier about the way that functions can be thought of as objects in themselves that can be manipulated and combined in a manner similar to ordinary algebraic symbols.

Combining functions

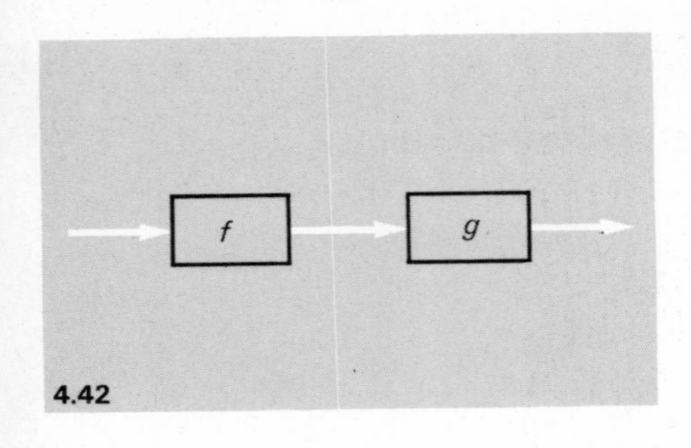

4.42

If you think of the interpretation of a function as a black box which is fed an input and produces an output, then there is no reason why we should not think of the possibility of using the output of the black box as the input to another one.

For example, suppose a motor car averages 30 miles to the gallon, then the function

$$f: x \to x/30$$

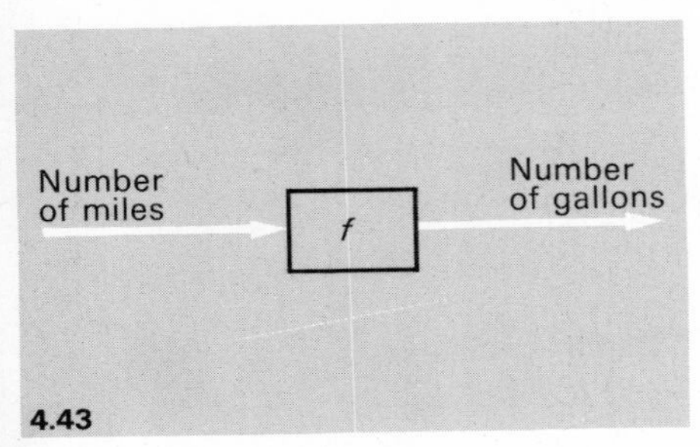

4.43

gives as its output the number of gallons required to travel any specified number of miles, x. So, to work out how many gallons you would need to travel 200 miles, say, you would just give x the value 200, producing the number $\frac{200}{30}$, which is $6\frac{2}{3}$. So $6\frac{2}{3}$ gallons corresponds to 200 miles, or

$$f: 200 \to 6\tfrac{2}{3}.$$

Now, suppose petrol costs 80p per gallon. The function

$$g: x \to 80x$$

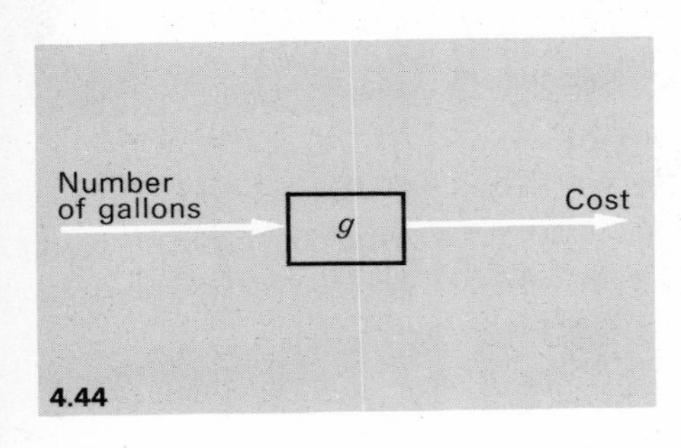

4.44

gives the cost of any specified number, x, of gallons. For example, if x is given the value $6\frac{2}{3}$, $80x$ becomes $533\frac{1}{3}$, so

$$g: 6\tfrac{2}{3} \to 533\tfrac{1}{3},$$

corresponding to the fact that $6\frac{2}{3}$ gallons cost $533\frac{1}{3}$ pence. (Notice by the way that we have used the same symbol, x, in the specification of f and of g – but remember that we are using x only as a tool, a 'stand-in' for any number; there is no reason to proliferate symbols by using a different one to specify g.)

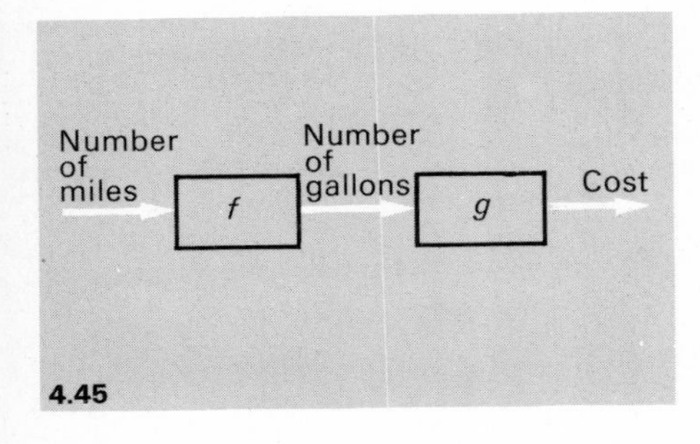

4.45

These two functions can be put together

and thought of as a single function to which we can input the number of miles and from which will be output the cost, directly.

So the two functions f and g are combined to give the single function h called the *composition* of f and g.

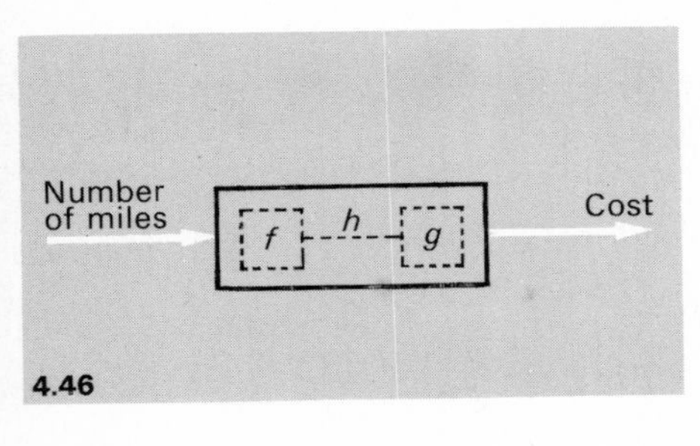

4.46

The composition of two functions, then, is the function obtained by using the functions successively, one after another. It is worth noting that it is not always possible to combine two functions in this way – the composition can be

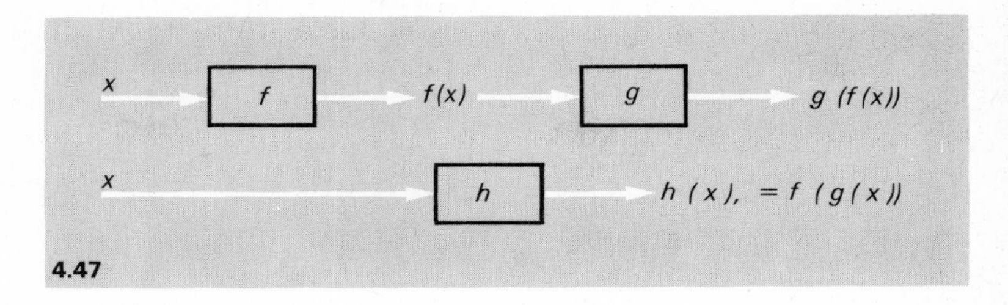

4.47

formed only if it makes sense to use the output of the first function as the input to the second.

For example, with our two functions describing the consumption and the cost of petrol, it would not make sense to combine them the other way round to the way we have done – think about it.

Symmetries again

Try cutting out a rectangle and place it on a piece of paper. Mark the corners–*A, B, C,* and *D* of the cut-out rectangle and on the paper below–then you can keep track of the movements.

Let us now see how this new idea works out with the functions describing the symmetries of the rectangle.

First of all what happens to the vertices of the rectangle if we use first the function f (reflection in the y-axis). If you feel happier about following the argument, it may help to cut out a rectangle from a piece of card and use that to go through the movements.

If we start like this

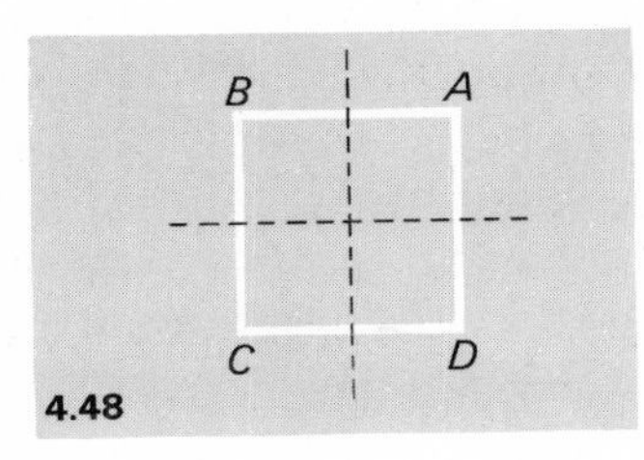

4.48

we get this.

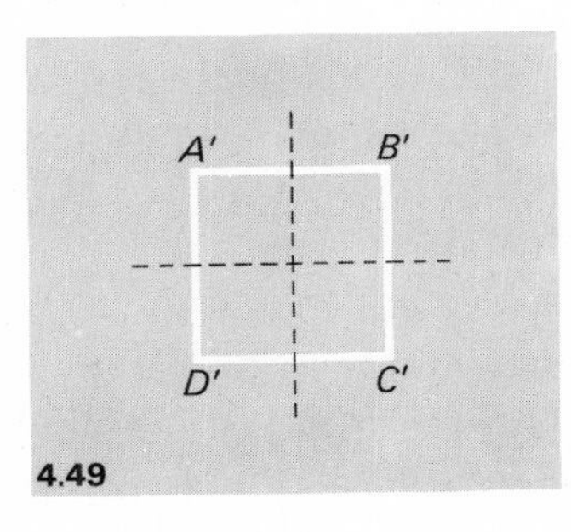

4.49

Now, starting from this position, what happens if we reflect in the x-axis? We get this.

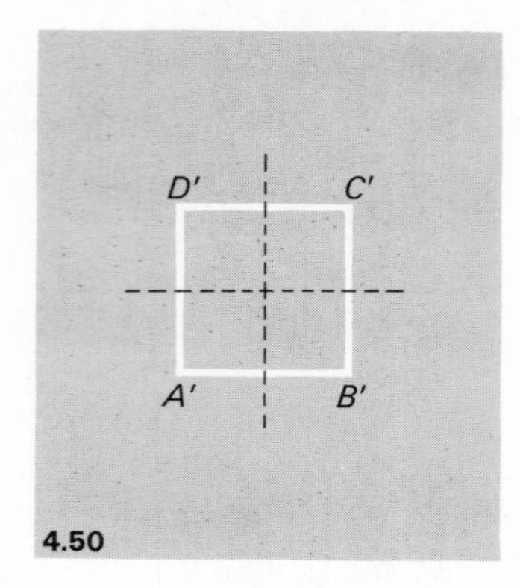

4.50

So the composite function, f followed by g takes this

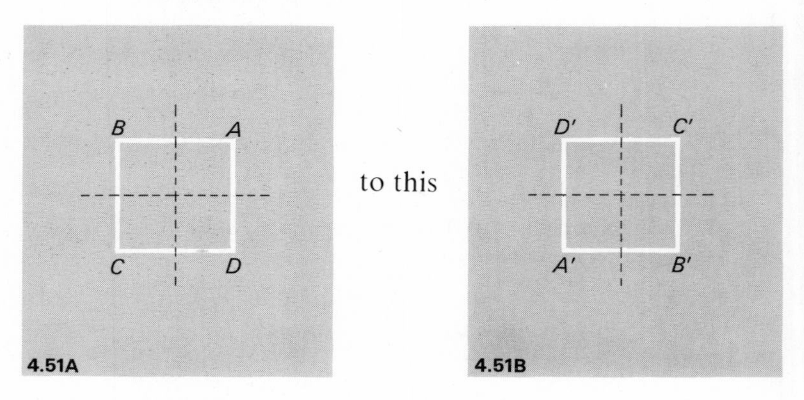

4.51A

4.51B

But the function that does just this, all in one go, is the function h, rotation about the centre through an angle of 180°.

It is usual to use a small circle, $\circ$, to stand for the composition of two functions, so the composition of f followed by g is $g \circ f$. (It is convention to write the first function used in the composition on the right, so $g \circ f$ means do f first and then g.) So we can write

$$g \circ f = h.$$

If you think for a few moments, you will realize that it is not really surprising that the composition of these two symmetry functions gives another one because the special feature of a symmetry function is that it leaves the rectangle invariant. No matter how many times you apply symmetry functions, the rectangle must remain invariant, so the total effect leaves it invariant and must correspond to a symmetry function.

This property of symmetry functions means that we can tabulate the results of combining pairs of such functions in a table like this:

$\circ$	e	f	g	h
e				
f				
g		h		
h				

The h is entered in 'row g' and 'column f' because $g \circ f = h$. Using this convention, the complete table looks like this; if you are interested, you may like to check the entries. By the way, you can now see a good reason for including e, the 'do nothing' function as a symmetry function. Without it, the table would have a lot of gaps in it.

$\circ$	e	f	g	h
e	e	f	g	h
f	f	e	h	g
g	g	h	e	f
h	h	g	f	e

This table gives a complete mathematical description of the symmetry properties of the rectangle. It is an example of a *mathematical model*. The purpose of such a model, as we have said, is to describe mathematically certain features of the subject of the investigation, whilst filtering out the features that are of no interest at the time. As before we see that not only does it enable us to concentrate on just those details that are of special interest but it can establish mathematical links between physical situations. For example, if you worked out a table of symmetries for another geometric figure and found it to be the same table as for the rectangle, then you would know that, as far as symmetry is concerned, you could regard your object as a rectangle – a simple and familiar shape that may be more easy to work with. Then again, you may have four functions arising from a completely non-geometric piece of work and you might find that they combined in just the same pattern as the symmetries of the rectangle. If that were the case, then you could talk of the non-geometric situation as possessing symmetry, even though it may have no direct geometric connotations.

The point is that sometimes a mathematical model is not only a way of using mathematical techniques to analyse a problem, it sometimes establishes a link between two situations which enables one physical context to be interpreted in another physical context – you may remember the example we had earlier about statements and switching circuits being linked by the common mathematical model of set algebra. In Chapter 6 we shall begin to develop one of the most famous and successful of all mathematical models – the use of mathematics to analyse motion.

5 Strange numbers

We have come some way from Chapter 1, where we first introduced symbols as ways of developing abstract arguments involving numbers. Since then we have used symbols in far more general contexts; but, however sophisticated mathematics may be, there is little doubt that numbers lie at its heart, and an understanding of numbers is essential to any appreciation of the subject. In this chapter we shall use some of the ideas we have developed so far to gain an insight into various types of numbers and the way they are related. Our general theme will be that of numbers as mathematical concepts invented to solve certain types of problems.

Natural numbers

Natural numbers are the counting numbers 1, 2, 3, ...

The most basic of all numbers are the ones we use for counting: 1, 2, 3, 4, ..., and so on. But have you ever thought how you might explain to somebody what you mean by these symbols? To explain them in terms of counting is fraught with danger because it is these numbers themselves that we use for counting. If you think back to the way you learnt about numbers, you will see that it was in fact a very indirect process. We learn to associate the number '3' with a set containing a certain number of objects – but not necessarily a particular set: any set with the same number of objects will do. A set of three oranges, three footballs, three television sets – with all of them we associate the number '3'. So a number is some undefined concept which we use in a well-understood way.

This is the way we are all taught to approach numbers and it is a perfectly acceptable attitude for the normal use we make of them. For mathematical use, however, a more precise point of view is desirable: not just for the sake of rigour, but because we need to extend our collection of numbers beyond those in ordinary use. We shall not go into all the philosophical problems, an idea of what is involved will be enough to help us explain some new types of numbers and give you an

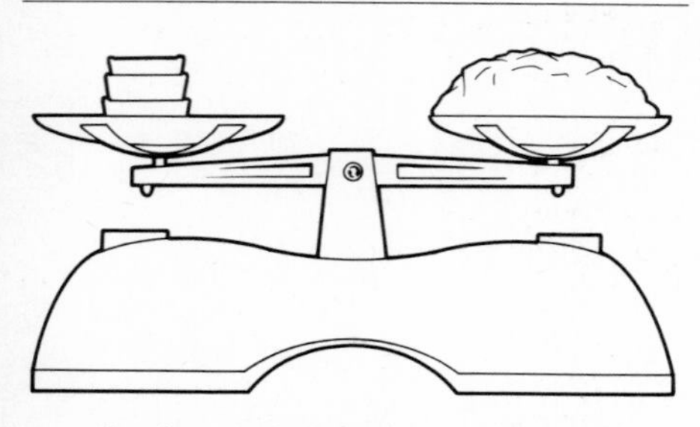

Scales weigh by *comparison*.

indication of the way mathematicians approach the problem of a rigorous definition.

It is really a question of making formal the way we approach numbers as children. As we said earlier, we associate the number '3' with something in common with many different sets. In other words rather than actually say what any particular number is, we find it easier to say when two numbers are the same. This is by no means unusual. The old type of scales tell you the weight of something merely by comparing it with a combination of standard weights. In much the same way, with numbers we observe a special relationship between certain sets – sets containing the same number of objects. But we began to see in the previous chapter that we have a good mathematical tool for investigating relations between sets in the idea of a *function*. Look back at p. 105 if you cannot quite remember what we mean by function.

Types of function

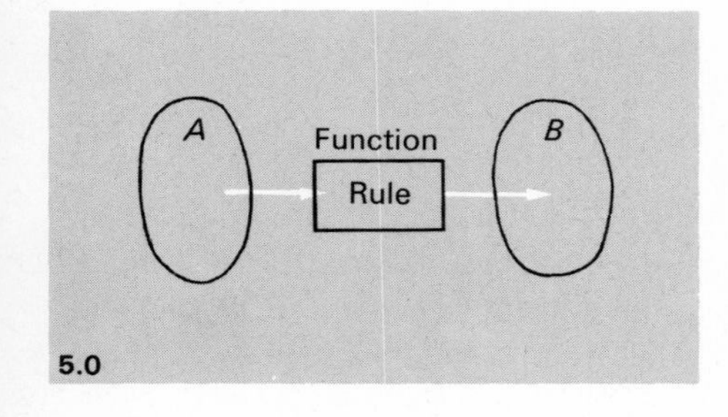

5.0

You may remember that we always require that a function associates an object in B with every object in A – in other words whatever object from A we feed in the function will give us an answer in B. We also insist that we get only one answer for each input. But even these restrictions still leave scope for various types of function. First of all, although every element of A must be assigned an element in B, it may be that not all the elements of B are used up in this way.

A B Set of all the images of elements of A

5.1 Some elements of B not assigned to elements of A

Into functions. The function that maps people to their ages maps the set of people *into* the set of numbers. Nobody gets mapped to 350, for example.

Such functions are said to map A into B, in contrast to those which exhaust all the elements of B – these map A *onto* B.

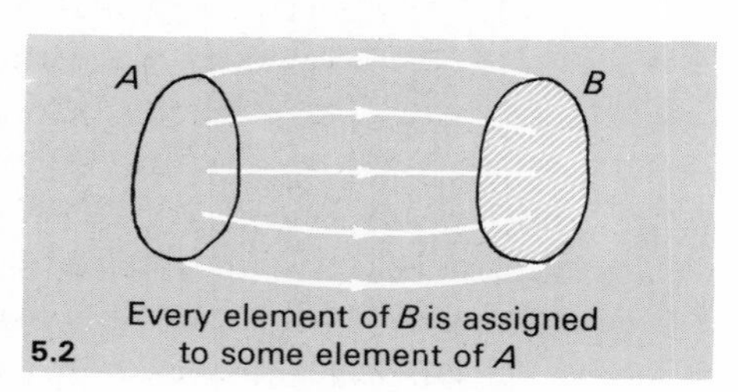

5.2 Every element of B is assigned to some element of A

So we can have two types of functions – *into* functions and *onto* functions. Within each of these two types we have two further classifications, illustrated like this.

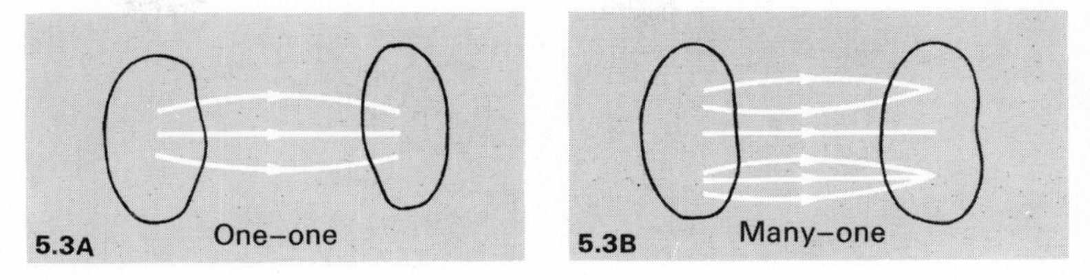

In the one–one case each element of B can be assigned to at most one element of A, whereas in the many–one case, some elements of B may be assigned to more than one element of A.

So we have four types of functions.

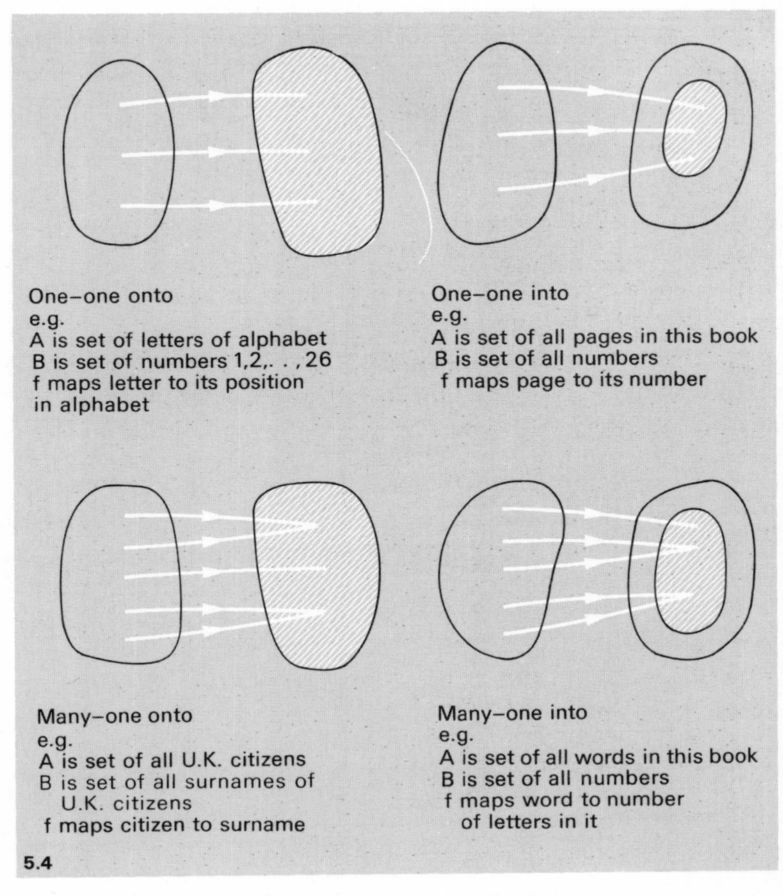

Our particular interest at the moment is the *one–one onto* type, because if a function of this type maps a set A to a set B, then it establishes a one-to-one correspondence between the two sets. In particular – and this is the important thing for us at the moment – the two sets must have the same number of elements. This is an observation which we can turn into a definition. First of all we note that 'number' is a concept that we associate with a set rather than with objects themselves and then we can say that two sets specify the same number if there is a one–one onto function between them. So, we are in a position to say when two numbers are the same, but we can't actually shed light on the meaning of the individual number 'in isolation'.

That is the basis of one way of establishing numbers and the arithmetic operations. In a sense the whole business is in the realm of philosophy. We shall not go into it here; we shall take the numbers 1, 2, 3, ... as understood, together with the combination of these numbers by addition and multiplication. But don't think that all our talk about functions was a waste of time – they have an important role to play in our discussion of other types of numbers. So let's now see what we understand by 'other types' of numbers.

New numbers

Why do we need numbers? Simply to describe and solve problems. But the range of problems that we can solve using just the numbers 1, 2, 3, ... is severely limited. Simple counting can be accommodated, even something rather more taxing like 'How many extra packets of sweets do I need if I want to give a packet to each of 10 children and I have only 3 packets in the cupboard?'

Algebraically, this problem of sharing sweets amounts to finding a number x such that $3+x = 10$, a particular case of the general problem of finding a number x such that

$$a+x = b,$$

where a and b are any specified numbers.

If b is greater than a, there is no difficulty in finding such an x. But in other cases, the numbers 1, 2, 3, ... do not supply a solution. For example, it is not possible to find a natural number x such that

$$7+x = 3.$$

To solve such problems we invent new numbers. A new number 0 is *defined* as the solution to

$$a+x = a,$$

and for every natural number, a, we invent a number, $-a$, which we *define* as the solution to

$$a+x = 0.$$

Integers
The set of whole numbers $\ldots -4, -3, -2, -1, 0, 1, 2, 3, \ldots$

You may think that this is a lot of fuss about negative numbers. After all, what with deficit financing, the government deals with negative numbers almost to the exclusion of positive numbers! But what we are after is to try to get a little below the surface, to get a feel for just how negative numbers fit in to the mathematical scheme of things. In this way, we can extend the natural numbers to the set, ... -6, -5, -4, -3, -2, -1, 0, 1, 2, ... This set is called the integers.

There are other methods of obtaining the integers but it is certainly the case that one of the characteristic properties of the integers is that in the set of integers the problem

$$a+x = b$$

is always guaranteed a solution. This equation cannot be solved when a is bigger than b with just positive numbers, but it can be if we have negative numbers available as well. It is this full set, of positive and negative numbers, that we call the integers.

In the same vein, we can make the next extension of our number system by considering the problem of finding solutions to equations of the form

$$a \times x = b.$$

If we work in the set of integers, we sometimes get a solution, sometimes not. For example, the equation

$$2 \times x = 10$$

has solution $x = 5$; but the equation

$$2 \times x = 7$$

has no solution from amongst the set of integers. As before, we extend the set of numbers by inventing new ones. The number $\frac{7}{2}$ is *defined* as the solution to the equation

$$2 \times x = 7,$$

and in general, the number $\frac{b}{a}$ is *defined* as the solution to the equation

$$a \times x = b\,.$$

So corresponding to every pair (a, b) of integers we get a number $\frac{`b'}{a}$. Numbers of this type, where b and a are integers, are called *rational numbers*, the term echoing the connection with ratios. For example, 1/4, 9/4, 6/3 are all examples of rational numbers. Rational numbers are fractions of the form $\frac{b}{a}$, where a and b are integers. You can remember the word 'rational' if you think of 'ratios'.

Rational numbers are fractions, of the form $\frac{a}{b}$, where a and b are whole numbers.

I am not trying here to indicate a method of *introducing* various types of numbers to children or at the other extreme to give a carefully rigorous mathematical development. Nevertheless, this approach *can* be formalized *and* you can probably recognize something in common with the way numbers are treated in schools. Remember, though, that all of these numbers are mathematical concepts – pure descriptions of ideas that have practical application beyond ordinary everyday arithmetic. This point of view has been expressed memorably by the philosopher A. N. Whitehead with whom Bertrand Russell collaborated to produce what was his greatest contribution to mathematics – *Principia mathematica* (published in three volumes between 1910 and 1913). Whitehead observed that

> The point about zero is that we do not use it in the operations of daily life. No one goes out to buy zero fish. It is in a way the most civilized of cardinals, and its use is only forced on us by the needs of civilized modes of thought.

So starting with the natural numbers and the operations of addition and multiplication we have introduced first the integers and then the rational numbers. If we are able to satisfy ourselves about the natural numbers and what it means to add and multiply them then an understanding of the other numbers and the arithmetic operations on them all follow.

The natural numbers, integers, and rational numbers are all familiar to you – even if the names aren't. The next step will take us into a realm far beyond the normal use of numbers. But it appears at first sight to be quite innocent: we consider the problem of finding a number x, such that

$$x^2 = a,$$

where a can be any specified rational number.

We have already seen, way back in Chapter 1, that the equation

$$x^2 = 2$$

raises problems in the sense that its 'solution', $x = \sqrt{2}$, cannot be expressed as a rational number – we referred then to $\sqrt{2}$ as being an 'irrational' number. We haven't yet got much of a feeling for what an irrational number is – or how many there are. Perhaps there are just one or two of them. We'll look at that problem shortly, but, before we embark on that, can we say now that once we can decide what we mean by an irrational number then we have exhausted all the types of numbers that we need?

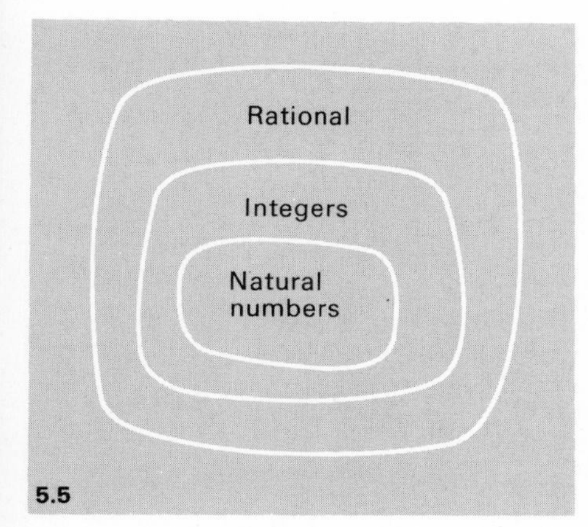

5.5

Taking stock

Just think for a moment what we have been doing so far in this chapter. We have started with the natural numbers and an assumption that we know how to add them and multiply them. Then we have extended our concept of number first to the integers and then to the rational numbers. But it has been a *genuine* extension: at each stage we have not lost what went before – we can think of the integers as a subset of the rational numbers and the natural numbers as a subset of the integers.

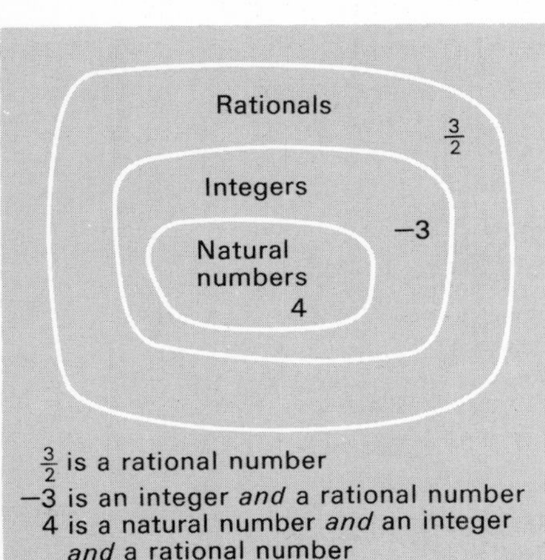

$\frac{3}{2}$ is a rational number
−3 is an integer *and* a rational number
4 is a natural number *and* an integer *and* a rational number

5.6

For example, $\frac{3}{2}$ is a rational number and so is $\frac{4}{1}$. But it is more usual to write $\frac{4}{1}$ just as 4, which is an integer. So some of the rational numbers are essentially just integers – and some of the integers, like 4 again, are for all practical purposes natural numbers.

Because of this relationship between these types of numbers, it is important that the arithmetic of these numbers carries through with each of the extensions. This means that the rules we have for multiplying and adding integers must be *deduced* from the definition of integers and the way we multiply and add the natural numbers. And the rules for multiplying and adding the rational numbers must be *deduced* from those for the integers. So if two rational numbers 'happen' to be integers and they are multiplied as rational numbers then we get the same answer as we would have got by multiplying them as integers. It is fairly easy to show that this works for the extensions that we have made so far, but we have now reached a stage where the extension is not so straightforward.

These last four sentences are important, but the ideas are not easy. Read them again to make sure you have got the general idea of what we are doing. Before you do so, it might help you if you see just a little more detail of what I mean by deducing multiplication of the rational numbers, for example, from multiplication of integers. Suppose x and y are rational numbers defined by

$$a \times x = b \text{ and } c \times y = d.$$

$\left(\text{In other words } x \text{ is } \frac{b}{a} \text{ and } y \text{ is } \frac{d}{c}.\right)$ Then, *just by using properties of the integers*, you can see that

$$ac \times xy = bd.$$

$\left(\text{In other words } xy \text{ is the number } \frac{bd}{ac}.\right)$ This gives the familiar rule

$$\frac{b}{a} \times \frac{d}{c} = \frac{bd}{ac}.$$

Remember our next step is to try to extend our number system so that we can solve equations of the form

$$x^2 = a,$$

where a is an integer. But one feature of the rational numbers is that if two negative numbers are multiplied together, the result is a positive number (see p. 44). This means that if the number a in the equation $x^2 = a$ is negative, then no matter how we extend the number system beyond the rational numbers, we cannot ensure a solution of the equation whilst retaining the arithmetic properties of the rational numbers. For example, there is no way of extending the rational numbers so that we can solve

$$x^2 = -4.$$

Certainly -2 will not do, because

$$-2 \times -2 = +4.$$

So our discussion now forks into two directions – we extend our number system to ensure the existence of solutions of $x^2 = a$ when a is not negative, and we invent a *new* number system to cope with equations $x^2 = a$ where a is negative. The completely new system we shall develop is the *complex-number system* and we shall look at complex numbers on p. 142. The extension of the rational numbers will be to append to them the irrational numbers and the full set of numbers thus obtained constitutes the *real-number system*, and we shall look at this first.

The real-number system

You may have sensed from the caution with which I am approaching the irrational numbers that they are by no means as straightforward as other numbers. This was hinted at in Chapter 1 and, in any case, is reflected in our experience at school – there is little attempt in school syllabuses to introduce irrational numbers in an explicit way. But what, you say, of $\sqrt{2}$ – that is something we use at school and haven't I been saying that $\sqrt{2}$ is an irrational number? The point is that at school we use only an approximation to $\sqrt{2}$. If you look in mathematical tables, or press the right button on a calculator, you will get something like 1.41425 – an ordinary decimal number. What I mean by 'ordinary' is that it terminates – this particular one has five numbers after the decimal point. Any such number, with a finite number of digits in it, is just a rational number. For example, 1.41425 is just the same as $\frac{141425}{100000}$, a rational number. And you can do the same sort of thing with any decimal that terminates. Now if you are performing a numerical calculation, the only type of number that you *can* handle is one with a finite number of decimal places – so the only numbers that you can use for calculations are rational numbers. The rational number 1.41425 is an *approximation* to the irrational number $\sqrt{2}$. So although every number is a purely mathematical invention, the irrational numbers are in some sense a step beyond even that – they are a purely mathematical invention to serve a mathematical need – to solve *mathematical* problems. Of course mathematics produces tremendous practical pay-offs, so the invention is not a selfish one even though it could be argued that the intrinsic value would be sufficient motive in itself.

There is something more in this idea of approximation than

is evident at first sight. In the first place, there are really many more problems other than solving $x^2 = a$ that cannot be solved using rational numbers alone and we shall see in a moment how the idea of approximation gives a way of defining other irrational numbers. But there is also something special about the way in which we use approximations to irrational numbers.

Explaining irrational numbers

We saw in Chapter 1 that, though you are forced to have to work always with approximations to irrational numbers, you *are* able to choose the accuracy of the approximation. The example we had in Chapter 1 will suffice to make the point. We saw there that the sequence of numbers

$$1, 2, 1.5, 1.666, 1.6, 1.625, 1.6154, \ldots,$$

calculated by working out the ratios of successive terms of the Fibonacci sequence

$$1, 1, 2, 3, 5, 8, 13, 21, 34, \ldots,$$

get closer and closer to the golden ratio.

The significant thing about the sequence of numbers

$$1, 2, 1.5, \ldots,$$

is that they get closer and closer together. More than that, by going far enough along in the sequence you can be sure that the terms have got as close together as you may wish. If somebody asked me to find a point in the sequence so that from that point, successive terms are never further apart than 0.0000001, say, then I know that I could do it.

The same idea applies to $\sqrt{2}$. Consider the sequence

$$a_1, a_2, a_3, a_4, \ldots,$$

where the numbers a_1, a_2, and so on are calculated using the formula

$$a_n = \tfrac{1}{2}\left(a_{n-1} + \frac{2}{a_{n-1}}\right)$$

starting with $a_1 = 1$.

This may look forbidding to some of you, but it isn't really. Let us feed some numbers in to show you how it goes.

We know that a_1 is 1. So we can calculate a_2 by putting $n = 2$ in the formula:

$$a_2 = \tfrac{1}{2}\left(a_1 + \frac{2}{a_1}\right)$$

with a_1 replaced by 1.

This gives us

$$\tfrac{1}{2}(1+2)$$

which is

$$1.5$$

The third term, a_3, is calculated by putting $n = 3$ in the formula:

$$a_3 = \tfrac{1}{2}\left(a_2 + \frac{2}{a_2}\right)$$

which is $\tfrac{1}{2}\left(1.5 + \frac{2}{1.5}\right)$ and turns out to be about 1.417. The fourth term works out to be about 1.4142. So the sequence of numbers turns out to be

$$1, 1.5, 1.417, 1.4142, \ldots,$$

and so on. Already you can see that the terms are beginning to bunch together.

Now *every* term in this sequence is a rational number, because all we are doing is adding, multiplying, or dividing rational numbers, but the 'number' around which they are bunching is $\sqrt{2}$, an irrational number! This is quite dramatic. We now have a link between rational and irrational numbers, even though not a direct link. It conveys a feeling that irrationals aren't purely intellectual things out of touch with physical reality. They are in a sense *accessible*. You can see this as follows. Let's suppose that a_n and a_{n-1} differ by an amount h, so that we can write

$$a_n - a_{n-1} = h.$$

This means that

$$a_n = a_{n-1} + h.$$

If we replace a_n by $a_{n-1} + h$ in the general formula

$$a_n = \tfrac{1}{2}\left(a_{n-1} + \frac{2}{a_{n-1}}\right)$$

we get

$$a_{n-1}+h=\tfrac{1}{2}\left(a_{n-1}+\frac{2}{a_{n-1}}\right).$$

If you do some work on this, you will find it works out to give

$$a_{n-1}^2+2ha_{n-1}-2=0.$$

Now look carefully at this. If a_n were close to a_{n-1} then, because $a_n=a_{n-1}+h$, h would be small. Can you see that smaller and smaller h becomes, the closer a_{n-1}^2 gets to 2, because the middle term, $2ha_{n-1}$ is getting very small? We haven't actually *proved* that the terms do get closer and closer together – so that h gets smaller and smaller, perhaps you can take it on trust that it can be proved. But we have shown now that if they do, then they get closer and closer to a number which when squared has the value 2 – a number which we would want to call the square root of 2.

If we insist on restricting our concept of number to the rational numbers, then something most strange happens. As we saw in Chapter 1, each of the numbers in the sequence can be represented by a point on the number line. The sequence of points clusters closer and closer about some point – the point we want to associate with $\sqrt{2}$.

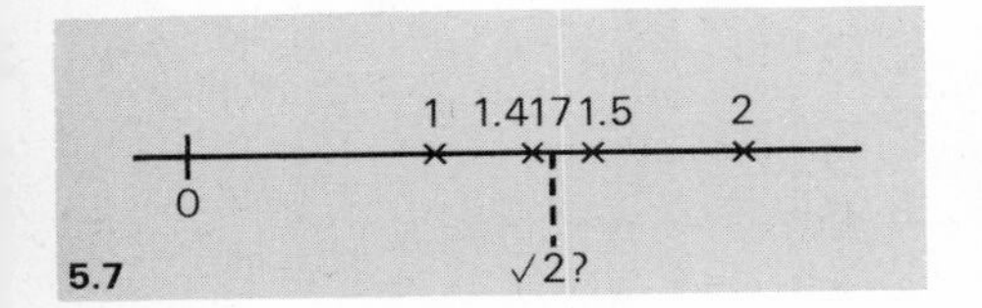

5.7

But since $\sqrt{2}$ is not a rational number there is some sort of 'gap' in the number line. This was well expressed by the German mathematician Richard Dedekind in 1872. He said

> The comparison ... of rational numbers with a straight line has led to the recognition of the existence of gaps, of a certain incompleteness or discontinuity of the former while we ascribe to the straight line completeness, absence of gaps, or continuity. In what, then, does this continuity exist?

Richard Dedekind (1831–1916) tackled the fundamental problem of explaining the nature of continuity in arithmetic terms. 'I succeeded November 24, 1858, and a few days afterward I communicated the results of my meditations to my dear friend Durège with whom I had a long and lively discussion.'

The word 'completeness' here is a key one, because if we allow the sequence

1, 1.5, 1.417, ...

to *define* a number (which we will call $\sqrt{2}$ of course) then in this way we can plug one of the gaps in the number line. And if we allow *any* such sequence, whose terms eventually approach arbitrarily close to each other, to *define* a number, then we can show that *all* of the gaps are plugged. This process is referred to as the *completion* of the number system and it introduces a whole collection of new numbers – called, of course, irrational numbers.

The complete collection

Now, perhaps, you can see why it took mathematicians so long to get their ideas straight about irrational numbers. This notion of a never-ending sequence of numbers is a much bigger step than the ones we had to take to move from the natural numbers to the integers or from the integers to the rationals. And there is a lot of detailed mathematical work to do – it is necessary to show how to add or multiply these strange new numbers, how to tell when they are equal and so on. We shall not go into that here, but in the next chapter we shall see a formal way of writing down what we mean by 'clustering closer and closer together' – one of the most important of all ideas in mathematics. For the moment, though, we want to look a little more closely at the nature of the number system that we have arrived at.

When a sequence of rational numbers clusters arbitrarily close, in the way that our sequence for $\sqrt{2}$ did, we say that it *converges*: the number around which the numbers are clustering (in our example it was $\sqrt{2}$) is called the *limit* of the sequence. These terms – *limit* and *convergence* – we shall define properly in Chapter 7, but we can say now that a complete system of numbers is defined by associating a number with the limit of every convergent sequence of rational numbers – 'complete' in the sense that it gives a one–one correspondence with the points on a line. Some of these limits will themselves be rational numbers – the other limits are the irrational numbers. The complete set of numbers is called the *real-number system.*

Algebraic equations

We were motivated earlier in this chapter to introduce irrational numbers to help us to solve problems such as

$$x^2 = a,$$

where a is not negative.

Clearly, there are a whole bunch of similar problems – solving

$$ax^3 + bx^2 + cx + d = 0$$

for example. By analogy with the way we extended the natural numbers to the integers and then the rational numbers, we might think of introducing the irrational numbers to ensure that we can at least in theory solve the general algebraic equation

$$a_n x^n + a_{n-1} x^{n-1} + a_{n-2} x^{n-2} + \dots + a_0 = 0,$$

where the numbers $a_0, a_1, \dots, a_n$ are any specified rational numbers and n can be any positive whole number. There are two difficulties here. In the first place we know that it is impossible! We have already seen that the equation

$$x^2 + 2 = 0$$

for example, cannot be satisfied by a number which obeys the normal laws of arithmetic (because it would require x^2 to be negative). We might try to get round this by thinking up some cunning ruse to exclude all these tricky cases and then say that all the equations that are left specify the complete real number system – rationals and irrationals.

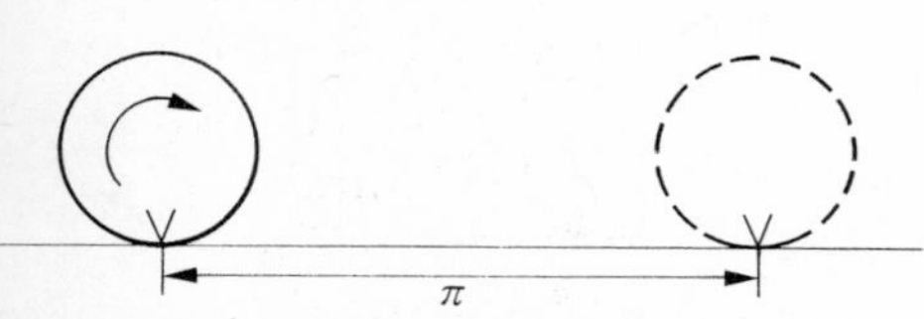

The circumference of a circle, radius r, is $2\pi r$. Roll a circle of radius 1 along a line and you can construct the point representing π.

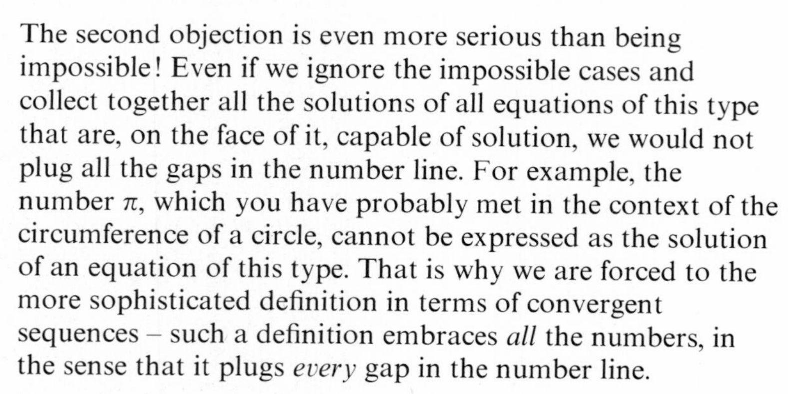

The second objection is even more serious than being impossible! Even if we ignore the impossible cases and collect together all the solutions of all equations of this type that are, on the face of it, capable of solution, we would not plug all the gaps in the number line. For example, the number π, which you have probably met in the context of the circumference of a circle, cannot be expressed as the solution of an equation of this type. That is why we are forced to the more sophisticated definition in terms of convergent sequences – such a definition embraces *all* the numbers, in the sense that it plugs *every* gap in the number line.

We cannot actually demonstrate here that the number π cannot be a solution of an algebraic equation, because the mathematics is rather advanced, but what we can do is to show that numbers like π must exist. That is to say, we can show that even if we include in our number system the solution of every possible equation of the form

$$a_n x^n + a_{n-1} x^{n-1} + \dots + a_0 = 0$$

then we still have gaps in the number line. A solution to an equation of this type is called an *algebraic number*. We shall be going into them in more detail shortly, but first let us try to see why algebraic numbers – numbers which satisfy equations like this – are not enough to specify the complete number system. We will see that this phenomenon can be demonstrated in a most ingenious way – which is all the more worth doing because it reveals some other interesting features of sets of numbers.

Counting sets

The way we shall tackle the problem is to try counting the real numbers and counting the algebraic numbers and in that way show that there are 'more' real numbers than algebraic numbers. That sounds a bit of a tall order. After all, if we try counting the integers we soon get into trouble – they go on for ever. The method we shall use is due to George Cantor who in the late nineteenth century was active in establishing the theory of sets and whose work has therefore penetrated almost every branch of mathematics. First, let's loosen up our ideas of counting.

We mentioned earlier that it is useful to think of counting in the same way that children do, as a pairing of the objects to be counted with some reference set such as the fingers. As the child gets older he finds he can dispense with his fingers and just counts out one, two, three . . . . He is again pairing off elements to be counted with elements in a reference set, but this time the set is more abstract – it is just the natural numbers – and he holds them in his head. In fact, this more mature approach is also more direct. The comparison, or pairing off is being made directly between the set to be counted and the natural numbers rather than the fingers which are a representation of the (first ten) natural numbers. As we saw at the beginning of this chapter this is just an intuitive expression of the fact that there is a one–one function from a subset of the natural numbers to the set in question. If the subset of the naturals numbers is the set 1, 2, 3, . . . , n, then the set has n elements.

Infinite sets

Let us try the same idea to count the set of positive even numbers. How do we count them? In the same way that we count anything else: we put them in one–one correspondence with a set of positive integers.

Positive even numbers:	2	4	6	8	…	24	…	$2n$
	↓	↓	↓	↓		↓		↓
Positive integers:	1	2	3	4	…	12	…	n

Can you see that the one–one correspondence is between the set of all even numbers and the *whole* set of positive integers? In other words the number of *even* positive numbers is the same as the number of *all* positive numbers! But the positive even numbers form a *subset* of the set of positive integers. So we have a situation where a set has the same number of elements as a subset of itself – something of a mental shock – especially if you have always thought that the number of integers is twice the number of even numbers!

The difficulty arises because the two sets do not have a finite number of elements – so what you mean by 'same number' is not quite as straightforward as you are used to. In fact, we could take as a *definition* of an infinite set one which can be put in one–one correspondence with a subset of itself. But the exciting thing is that this extends our notion of counting. It does not enable us actually to *count* infinite sets – that could hardly be expected – but it does enable us to compare them and in a very sensible way say what we mean by an infinite set having 'more' elements than another.

Intuitively an infinite set is an infinite set and that is that – it does not seem to make much sense to talk of different types of infinity – to talk of one infinite set having 'more' elements than another. But we generalize the notion of 'more' in a way that we have seen other concepts generalized: we express it in a familiar context in a manner that makes sense in the general context. We can express the familiar idea of a finite set having 'more' elements than another finite set in an abstract way using functions. If there is a one–one function from a set A *into* a set B then there must be some elements in B left out of the pairing of elements of A with elements of B: the set B has more elements than the set A.

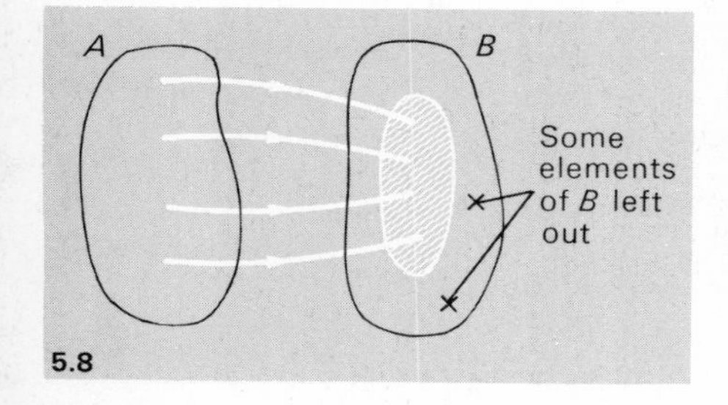

5.8

This is an idea that can be generalized to infinite sets. If A and B are infinite sets and there is a one–one function from A *into* B, then in a sense we can say that B has 'more' elements than A. For ease of reference we introduce the term *countable*: any set B for which there is a one–one function from the natural numbers, or a subset of the natural numbers, to B is called *countable*. For finite sets, we would have a finite subset of the natural numbers, corresponding to our usual concept of counting.

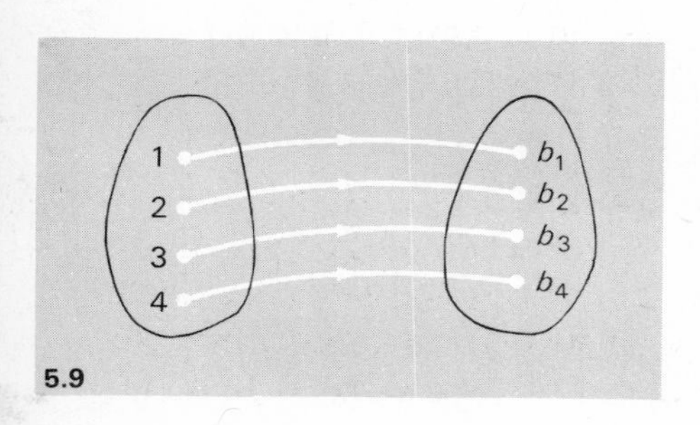

5.9

For infinite sets, the idea corresponds to a 'matching up' with the natural numbers.

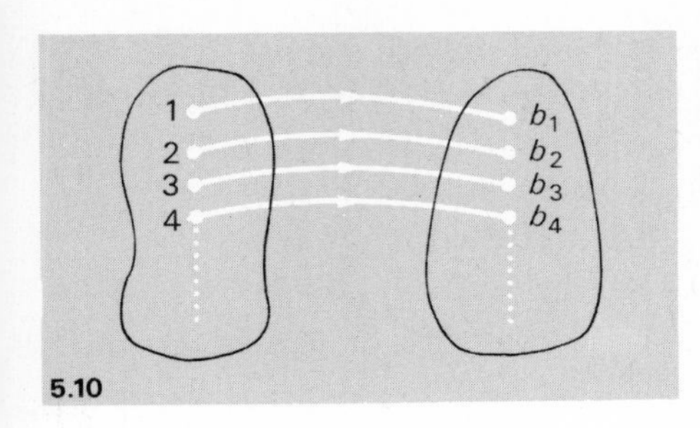

5.10

Although you cannot actually *count* the number of elements in B, the term 'countable' is used in an attempt to convey the idea that B does at least match up with the counting numbers. If on the other hand you can show that after such a matching up, there are always some elements of B left out, then B is *uncountable.*

Uncountable sets must, then, have an essentially different character to countable ones and our next job is to demonstrate an uncountable set: and the one we choose is the set of real numbers!

An argument by contradiction

Argument by contradiction

Assume statement false

↓

Deduce contradiction

↓

Conclude statement true

We are going to use an argument you have seen earlier – the method of contradiction (p. 22). We shall assume that the real numbers are countable and *prove* that this leads to a contradiction.

Remember that the set of real numbers is the complete set that we built up – from the natural numbers to the integers to the rational numbers and then to the irrational numbers.

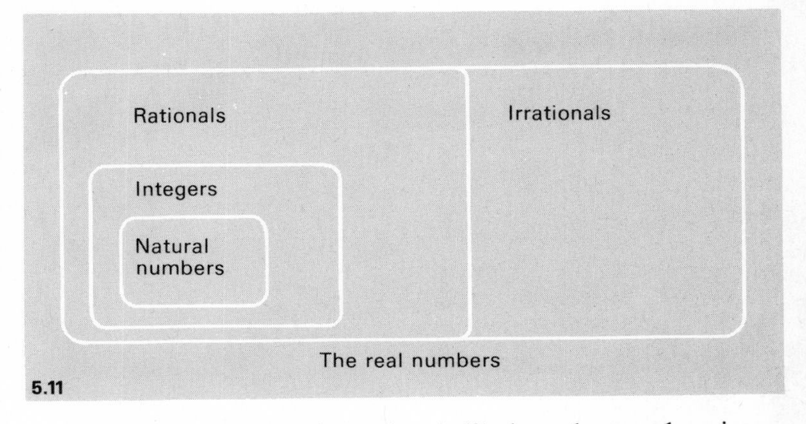

5.11

A word of warning! This section is likely to be tough going. But it is, after all, an important and rather difficult piece of mathematics. So if you can manage to tackle it, you can congratulate yourself – you will have achieved something worthwhile.

Let us suppose, then, that we *can* find a one–one function from the natural numbers *onto* the real numbers – this is equivalent to saying that the set of real numbers is countable. Now any real number can be expressed in a decimal form, for example $1/8 = 0.125$, $1/3 = 0.3333\ldots$ (where the 3s go on for ever). This is the representation that we are going to use in our argument. We have not gone into decimal representation of numbers but there is not much you need to know here apart from the fact that it can always be done. Perhaps you can take that on trust. As you might expect, each real number has a unique decimal representation; you cannot find two decimal forms for the same number. (The only exception is that the numbers 0.9999 ... and 1.0000 ... are the same. Remember, that the 9s go on 'for ever'. It is a technical point that you need not worry about, and we can get round it just by agreeing not to allow an infinite string of 9s.)

So let us suppose all of the *real numbers* to be written in decimal form, that they are countable and that they can therefore be put into one–one correspondence with the numbers 1, 2, 3, ... Suppose the numbers corresponding to 1, 2, and 3, go like this:

$$1 \rightarrow a_0 \cdot a_1 a_2 a_3 a_4 \ldots$$

$$2 \rightarrow b_0 \cdot b_1 b_2 b_3 b_4 \ldots$$

$$3 \rightarrow c_0 \cdot c_1 c_2 c_3 c_4 \ldots$$

and so on.

This all looks very difficult, but all the *a*s, *b*s, and *c*s just stand for ordinary numbers. We have to use suffices because we would soon run out of letters otherwise and in any case the suffix is a handy way of noting which decimal place we are talking about.

Remember that we are trying to show that the set of real numbers is not countable and our plan is to show that in our listing above we could not have included every number. We shall do this by *constructing* a number that is not on the list. Because we have written down the list using symbols, rather than a particular list, with actual numbers, our argument will apply to *any* listing that anyone could possibly propose. This is important, because we are trying to show that no complete listing could exist. Let's write

$$x = x_0 \cdot x_1 x_2 x_3 x_4 \ldots.$$

This number, x, is going to be a number *not* on the list. We can construct it like this. To choose x_0, we look at a_0. Whatever a_0 is we choose x_0 to be different. This ensures that x is different from the first number in the list.

To choose x_1 we look at b_1. Whatever b_1 is we choose x_1 to be different. This procedure ensures that x is different from the second number in the list. We can use a similar device to ensure that the second decimal place of x is different from c_2, the second decimal place of the third number on the list. And we can make the kth decimal place of x to be different from the kth decimal place of the $(k+1)$th number in the list.

The set of real numbers is *not* countable.

The number x constructed in this way is different from each of the numbers in our list because whatever number we choose from the list, it differs from x in at least one decimal place. This contradicts our original suggestion that *every* number appears in the list somewhere. The inescapable conclusion is that the set of real numbers is not countable.

Algebraic numbers

Let's just recall why we set out on this business about countable sets. Remember, we wanted to show that not every real number can be regarded as a solution of an equation of the form

$$a_n x^n + a_{n-1} x^{n-1} + \ldots + a_0 = 0.$$

The set of algebraic numbers *is* countable.

What we are going to do is to show that the set of all real numbers that are solutions of equations like this form a countable set and that because the real numbers are *not* countable, there must therefore be some real numbers outside this set.

There must be some real numbers that are not algebraic numbers.

First remember that we have given this special set of real numbers a name. An equation of the form

$$a_n x^n + a_{n-1} x^{n-1} + \ldots + a_0 = 0,$$

where a_0, a_1, ... up to a_n are integers is called an *algebraic equation*. Any real number that satisfies an algebraic equation is called an *algebraic number*. You may have noticed that we now specify the numbers a_n, a_{n-1}, ..., a_0 as integers, whereas previously we required them to be rational numbers. This is just a slight simplification, not a significant change, because the rational numbers can all be changed to integers by multiplying right through the equation by the appropriate integer. For example, the equation

$$\tfrac{2}{3}x^2 + \tfrac{1}{5} = 0$$

is equivalent to the equation obtained by multiplying right through by 15:

$$10x^2 + 3 = 0.$$

Restricting the numbers a_0, a_1, ... to be integers just makes the argument a little easier.

The algebraic numbers are a very big class of numbers, for example it includes all the rational numbers; the number $\frac{b}{a}$ satisfies the algebraic equation

$$ax - b = 0.$$

We have also seen that it includes some irrational numbers, for example $\sqrt{2}$ satisfies the algebraic equation

$$x^2 - 2 = 0.$$

But as we remarked earlier, algebraic numbers do not account for the complete real-number system. We have seen that the real numbers are not a countable set: we are now going to show that the algebraic numbers *are* countable, and so there must be some real numbers that are not algebraic. We do this in a rather cunning way.

Algebraic numbers are countable

First, we need to be able to refer to a general algebraic expression, so instead of looking at particular examples, like $x^2 + 2$, $x^3 - 3x + 6$, and so on, we use *letters* to stand for general numbers. So when we consider expressions like

$$a_0 + a_1 x + a_2 x^2 + \ldots + a_n x^n,$$

a_0, a_1, a_2 ..., a_n are just 'stand-ins' for any integer you like to think of.

Each algebraic expression,

$$f(x) = a_0 + a_1 x + a_2 x^2 + \ldots + a_n x^n,$$

where not all of the numbers a_0, a_1, ..., a_n are zero, is specified by the numbers a_0, a_1, a_2, ..., a_n. Now we form the number

$$l = |a_0| + |a_1| + |a_2| + \ldots + |a_n| + n.$$

(Remember from p. 44 that $|a|$, the modulus of a number a is just its numerical value, so $|-3| = 3$, for example but $|7| = 7$, and so on.) We can calculate l for any of our algebraic expressions: for example if $f(x) = 2-3x$, $l = |2|+|-3|+1 = 6$. Since the as are all integers, l must also be an integer; what is more it is positive, because we have taken the modulus of each of the as. So l could turn out to be any of the numbers 1, 2, 3, ... Can you see that for any particular value of l, there can be only a finite numbers of expressions $f(x)$? For example suppose l were 3. There is only a finite number of ways of choosing positive numbers to add to 3. And n cannot be greater than 3, because of the way l is calculated – so the number of terms in the expression $f(x)$ is limited. Let's look at the first few possibilities. If $l = 1$, n must be 0, because not all the as can be zero, so the only possible expressions are

$$1 \quad (a_0 = 1, \text{ the rest zero})$$

and $-1 \quad (a_0 = -1, \text{ the rest zero})$.

If $l = 2$, n cannot be bigger than 1, and the only possible expressions are

$$f(x) = 2 \quad (n = 0,\, a_0 = 2),$$
$$f(x) = -2 \quad (n = 0,\, a_0 = -2),$$
$$f(x) = x \quad (n = 1,\, a_0 = 0,\, a_1 = 1),$$
$$f(x) = -x \quad (n = 1,\, a_0 = 0,\, a_1 = -1).$$

You may like to try working out the possibilities for $l = 3$, but in any case you should be convinced that for each value of l these are only a finite number of possibilities for $f(x)$. So we can list all the possibilities – say for $l = 1$ we get $f_1(x)$ and $f_2(x)$ and for $l = 2$ we get $f_3(x), f_4(x), f_5(x), f_6(x)$, and so on. This gives us a sequence of expressions

$$f_1(x), f_2(x), f_3(x), \ldots.$$

Now, although we have not proved it anywhere, it is a fact that an algebraic equation of degree n (that is to say where the highest power of x is x^n – so that $x^3+3x+1 = 0$, for example, is of degree 3) has *at most* n real numbers as solutions (although not necessarily as many as n). So any one of the expressions $f_k(x)$ can specify only a finite number of algebraic numbers, as solutions of $f_k(x) = 0$. So for *each* $f_k(x)$ we can write down a finite sequence of algebraic numbers.

In this way we get a whole sequence of algebraic numbers, like this:

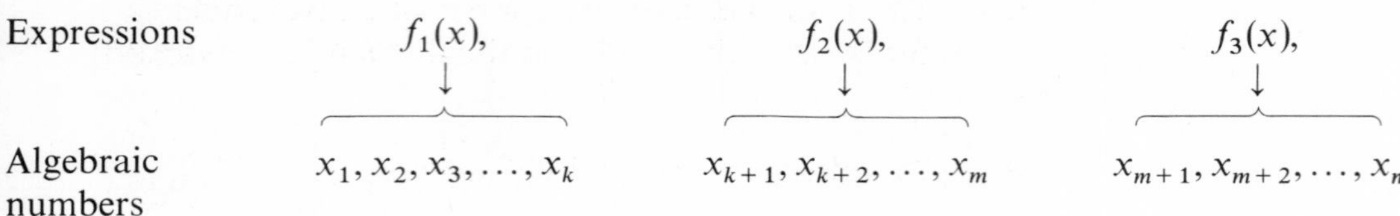

Every algebraic number will appear in this list – because it covers every real-number solution of every algebraic equation and so we can use it to set up a one–one function from the set of natural numbers to the set of algebraic numbers. A first shot at this would be to associate with the number m the mth algebraic number along the list. But there is a slight technical difficulty – there are many repetitions in the list of algebraic numbers. So we modify the function slightly and associate with the number m, the mth *distinct* algebraic number in the list. In other words, you work along the list rejecting any number that you have already met in the list. This establishes a one–one correspondence between the natural numbers and the algebraic numbers and thus demonstrates that the set of algebraic numbers is countable. Since as we have seen, the real numbers are not countable, there must exist real numbers that are not algebraic. Such numbers are called *transcendental numbers.*

Now that proof was *by no means* easy. But don't worry, it is not vital that you follow every detail. But the general drift of the proof is interesting.

1. We associated with each algebraic equation a positive number. We know that we can 'count' the positive numbers.

2. Associated with each positive integer there can be only a *finite number* of algebraic equations. So corresponding to each positive integer we can *count* the number of equations.

3. We then noted that every algebraic equation specifies a *finite number* of algebraic numbers. So the algebraic numbers specified by any particular algebraic equation can be counted.

4. Putting all this together we can see that the whole set of algebraic numbers can be 'counted'.

If you did find this proof a little strange and rather difficult, it is hardly surprising, because it is a type of argument that rarely crops up in elementary mathematics. Nevertheless, it has some features that arise continually in more advanced work.

In the first place, it is an example of an *existence theorem*: we have not actually found any transcendental numbers, we have simply demonstrated that they must exist. And, further, the method of proving the existence theorem involved as a sort of subplot another existence theorem! We did not actually find an explicit function which told us where any particular algebraic number was along a list of algebraic numbers. For example, we still haven't shown how to assign a natural number to $\sqrt{2}$. All we have done is to demonstrate that it is possible to find such a function, to perform such a listing – and that is all we need to do.

It is one thing to know that transcendental numbers exist, but it is quite another to find one. To prove a number to be transcendental you have to prove that no algebraic equation exists of which the number in question is a solution. And there are an awful lot of equations to try. One such number is π, as we mentioned earlier in this chapter, but it was not *proved* to be transcendental until as comparatively recently as 1882. The proof involves some rather advanced mathematics and so we cannot go into it here. Some other, rather weird transcendental numbers had been discovered before this proof was published, but nevertheless the proof was something of a breakthrough because it settled a problem outstanding since the time of the Greeks.

This is the problem of squaring the circle – using only a straight-edge and a compass to construct a square whose area is the same as the area of a given circle. If the circle has radius 1, then its area is just π because the area of a circle is π times the square of the radius. So our problem is to construct a square with area equal to π. The side of such a square would have to be $\sqrt{\pi}$. The Greeks knew how to construct a line of length $\sqrt{\pi}$ from a given line of length π, by using Pythagoras' theorem and constructing a right-angled triangle.

What worried the Greeks was how to construct the line of length π. Any line that can be constructed using just a straight edge and compass has a length that can be expressed in terms of integers using the operations $+$, $-$, $\times$, $\div$, and $\sqrt{}$, and so it must be algebraic. So when π was shown not to be algebraic the Greek's problem was finally settled – it was impossible!

So let's summarize what we have learnt about the real-number system. The most elementary types are:

natural numbers: the counting numbers 1, 2, 3, 4, ... ;

the *integers*: 0, 1, -1, 2, -2, 3, -3, ... ; and

the *rational numbers*: fractions, $\frac{a}{b}$ where a and b are integers.

Sequences of rational numbers, $x_1, x_2, x_3, \ldots$ which converge – a term to be defined properly in the next chapter – specify numbers as the limits of such sequences (the limit being the number about which the numbers in the sequence cluster closer and closer – again a concept we will define properly in the next chapter). The complete set of all possible numbers specified in this way is the set of *real numbers*.

The complex-number system

Although our main concern in this book is with mathematics involving only the set of real numbers we include a mention of the *complex numbers* for three reasons. In the first place, it will tie up one or two loose ends from our discussion so far. Secondly, it will make the point that there are more numbers, beyond the real number system, and thus explain why we have to use the term 'real' number system, rather than simply 'the' number system – although, as we shall point out, the adjective 'real' will still not be explained in an entirely satisfactory way. Thirdly, the complex number system is an example of a pure creation of the intellect and one that, for all its elegance as a concept in its own right, has played a vital role in the application of mathematics to the solution of practical problems.

There are many ways of introducing complex numbers; this is a reflection of their ubiquitous character. In the context of the earlier part of this chapter the most relevant approach is to refer again to the solution of equations.

You may recall that, on p. 127, we reached a point in the development of the real numbers where the chain of development branched. We were considering solutions of algebraic equations, of the form

$$a_n x^n + a_{n-1} x^{n-1} + \ldots + a_1 x + a_0 = 0$$

as a way of extending the number system beyond the rational numbers. We noted then that not every equation of this type had solutions in the set of real numbers. The easiest such equation is

$$x^2 + 1 = 0 .$$

We pointed out that if we want to preserve the rules of arithmetic as they apply to 'normal' numbers then we cannot extend these 'normal' numbers so as to ensure that this equation has a solution. At the time we restricted our attention to those equations that do have solutions which obey the 'usual' laws of arithmetic. Now, we are going on a flight of fancy – to see what happens if we abandon the restrictions dictated by convention.

The first thing we have to do is to introduce some notation – let us denote the 'solution' of the equation $x^2 + 1 = 0$ by the symbol i. This is a new use of symbols in this book. So far we have been using letters to stand for 'conventional' numbers – conventional in the sense of representing a physical quantity such as length. The use of the symbol i is not like this at all, because it has the strange property that when it is 'multiplied' by itself,

$$\mathrm{i} \times \mathrm{i} = -1 .$$

We use this symbol to construct a new class of objects – called complex numbers – like this.

A complex number has the form

$$a + b\mathrm{i} ,$$

where a and b are real numbers, and i satisfies the equation $x^2 + 1 = 0$. They are added and multiplied in a similar fashion to the ordinary algebraic operations:

$$(a + \mathrm{i}b) + (c + \mathrm{i}d) = (a + c) + (b + d)\mathrm{i}$$

$$(a + b\mathrm{i}) \times (c + d\mathrm{i}) = a \times c + b\mathrm{i} \times c + a \times d\mathrm{i} + b\mathrm{i} \times d\mathrm{i} .$$

You can see that addition of two numbers produces another complex number – for $a + c$ and $b + d$ are just real numbers, and so the answer is of the prescribed form: real number plus real number × i. By collecting together terms in the multiplication you can see that again you get a complex number, although at first sight it may not seem so because of the term $bi \times di$. But if we rewrite this as $b \times d \times i \times i$ and recall that i is a solution of $x^2 + 1 = 0$, then we can replace $i \times i$ by -1 – because we 'know' that $i \times i + 1$ must be 0. This gives

$$(a + bi) \times (c + di) = (ac - bd) + (ad + bc)i$$

which is of the form of a complex number:

Real number + Real number × i.

This procedure of defining the complex-number system may seem to you to beg a lot of questions. For example, how do we know that we can rearrange $bi \times di$ to $b \times d \times i \times i$? If the rules are going out of the window – so that we can propose the square of a quantity to be negative why should other laws of algebra stand? The only real answer to that is that we abandon only those laws that we have had to and preserve the rest. There is no justification beyond that except the overiding one that we happen to end up with an extremely useful mathematical tool.

This method of defining complex numbers is the historical one – it was how they were first constructed. There have since been many alternative approaches and many of them give sound reasons why the complex numbers should be constructed in the way they are. But all these reasons result from the benefits of hindsight: there is no denying the imaginative insight of the original creation of the ideas, which went beyond cold logic.

It is difficult to give convincing examples of the sheer power of complex numbers – their potency is revealed most markedly in more advanced mathematics – but highly practical mathematics – with applications as far ranging as from electronics to aeronautical engineering. But one quite impressive fact is available to us now. Do you remember, when we were looking at the equation

$$a_n x^n + a_{n-1} x^{n-1} + \dots + a_0 = 0,$$

The potential of complex numbers was probably first realised by the great German mathematician Gauss (1777–1855). It was he who first proved that every algebraic equation of degree n has n solutions in the set of complex numbers.

---

we remarked that not all of its solutions were real numbers? So although the consideration of this equation led us to the development of more real numbers, the real-number system was still not 'rich' enough to ensure the solution of every equation of this type. It turns out that the set of complex numbers meets this requirement very nicely. If we are allowed to choose numbers from the set of complex numbers, then *any* equation of the form

$$a_n x^n + \dots + a_0 = 0,$$

has a solution. More than that, it has exactly n solutions.

So, just by considering a solution of one equation of this type:

$$x^2 + 1 = 0,$$

we are able to construct a system which contains solutions of *all* such equations! But is it not absolutely fascinating to think that you have to open your mind to accept thinking in terms of *pairs* of numbers to solve them?

Representing complex numbers geometrically

One other example of the richness and extraordinary fascination of the complex numbers arises from our ability to represent them geometrically.

5.12

We have used the representation of real numbers as points on a number line.

Any complex number $a + bi$ is specified by a *pair* of real numbers (a, b). So it can be represented by a point on a plane, using *two* number lines and exploiting the usual method of plotting points using coordinates.

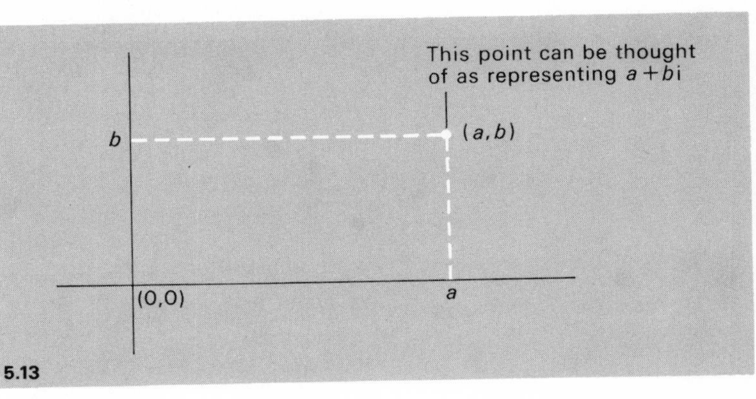

5.13

An algebra of pairs

Before we develop this geometric representation further, there is an interesting aspect of the observation we just made about the fact that a complex number is specified by two real numbers.

With an eye on the way we constructed the complex numbers, we can develop an alternative construction which avoids the introduction of the symbol i, which may have worried you a little.

We postulate a system as follows. We take the set of all pairs of real numbers, like (a, b). There is no difficulty in this because we already know that there is a correspondence between points and pairs, and so this present set is just an algebraic model of the set of all points in a plane. But the next step *is* new: we propose two methods of combining these pairs. The first we call 'addition of pairs', and we define it like this:

$$(a, b)+(c, d) = (a+c, b+d).$$

The second we call 'multiplication of pairs', and this one works like this:

$$(a, b)\times(c, d) = (ac-bd, ad+bc).$$

This rather bizarre definition looks far more sane when you compare it with the way we have already multiplied complex numbers – on p. 144:

$$(a+b\mathrm{i})\times(c+d\mathrm{i}) = (ac-bd)+(ad+bc)\mathrm{i}.$$

Can you see that the set of pairs with these methods of addition and multiplication are a model of our complex-number system? What is more the symbol i does not appear in either of the rules ('addition' and 'multiplication') for combining pairs. So the set of pairs with these methods of combining them form a model constructed without any reference to the mysterious solution of $x^2+1 = 0$. It is specified *entirely* in terms of the real numbers and the conventional methods of multiplying and adding real numbers.

But there is something else; this new system can be seen as a genuine extension of the real numbers because amongst this set of pairs are the pairs of the form $(a, 0)$ where the second number in the pair is just zero. If you put both b and d equal to zero, then our method of adding and multiplying *pairs* gives

$$(a, 0)+(c, 0) = (a+c, 0)$$

and $(a, 0)\times(c, 0) = (ac, 0)$.

So if we associate these particular pairs with real numbers, so that $(a, 0)$ is just a strange way of writing the single number a, then these two combinations are exactly analogous to the ordinary arithmetic of real numbers, because $(a+c, 0)$ and $(ac, 0)$ are just ways of writing $a+c$ and ac respectively. Can you see, then, how we can think of the complex numbers as an extension, albeit a rather dramatic one, of the real numbers, because in a genuine sense, the real-number system sits inside the complex number system. You can just think of a real number a as a complex number $(a, 0)$ – with the second number in the pair equal to zero. Notice also that our multiplication of pairs leads to the result that

$$(0, 1)\times(0, 1) = (-1, 0) \quad \text{(compare this with } \mathrm{i}\times\mathrm{i} = -1)$$

by directly substituting zero for a and c and 1 for b and d. The equation

$$x^2+1 = 0$$

written in terms of pairs looks like this:

$$x^2+(1, 0) = (0, 0) \qquad \text{(compare with } x^2+1 = 0)$$

and if we replace x by the pair $(0, 1)$ we get for the left-hand side

$$(0, 1)\times(0, 1)+(1, 0) \qquad \text{(compare with } \mathrm{i}\times\mathrm{i}+1)$$

which we can write as

$$(-1, 0)+(1, 0) \qquad \text{(compare with } -1+1)$$

which amounts to

$$(0, 0)$$

So $(0, 1)$ is a solution of this equation when you operate in the system of pairs.

An arithmetic of points

Let's explain these ideas just a little further. We know that there is a correspondence between points in a plane and pairs of real numbers. Now we have two methods of combining pairs of numbers – an arithmetic of pairs – combining two

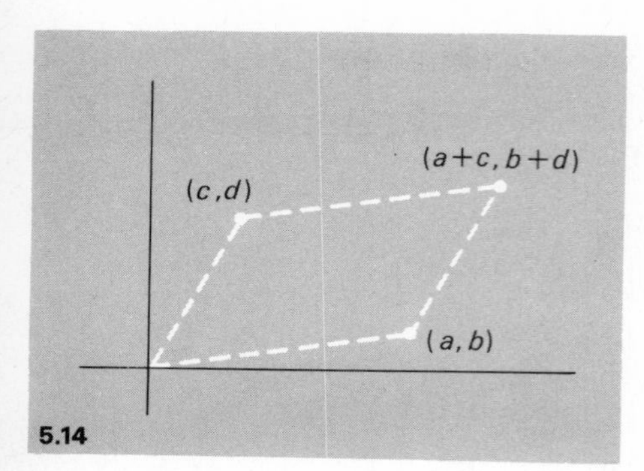

5.14

pairs (either by addition or by multiplication) produces another pair. But these can be interpreted as methods of combining points to produce other points – an arithmetic of points. Addition is easy to interpret geometrically. It turns out that $(a+c, b+d)$ is the vertex of the parallelogram whose other three vertices are at $(0, 0)$, (a, b), and (c, d).

An alternative, and rather interesting way of looking at it is like this. A pair (x, y) can be thought of as an *instruction* to move x units horizontally and y units vertically. Starting at $(0, 0)$ this instruction takes us to (x, y) of course. But starting at (a, b) it takes us to $(a+x, b+y)$.

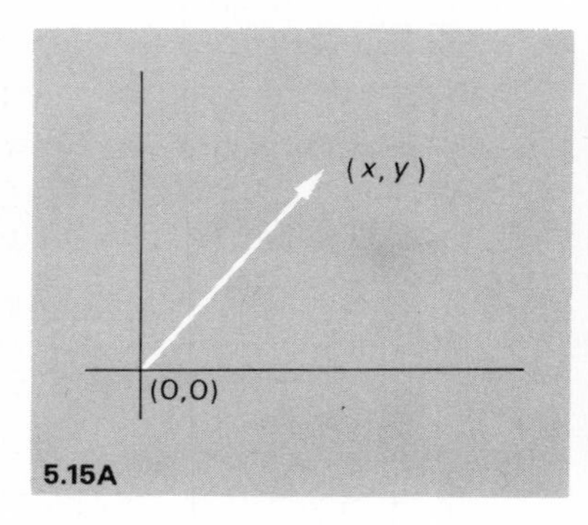

5.15A

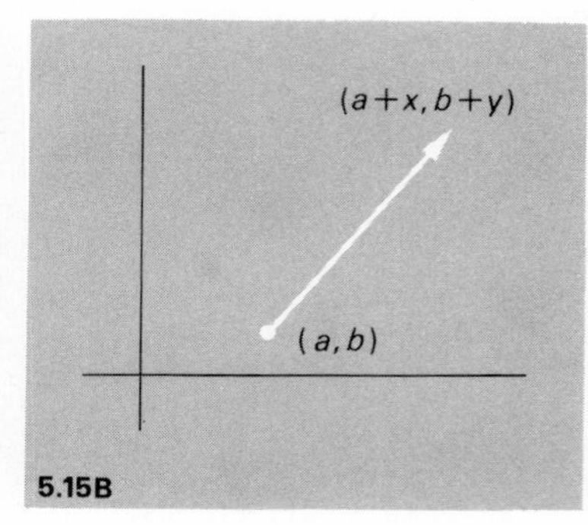

5.15B

So the sum,

$$(a, b)+(c, d) = (a+c, b+d),$$

can be interpreted as a combination of instructions – starting from the origin go first a units horizontally and b units vertically and then follow this by moving c units horizontally and d units vertically.

5.16

The product,

$$(a, b)+(c, d) = (ac-bd, ad+bc),$$

is rather more difficult to interpret geometrically because it requires some knowledge of trigonometry, but the result is very interesting. An alternative method of specifying a point, P, in a plane is to specify its distance from the origin, O, and the angle that OP makes with the horizontal axis.

5.17

So, since a point can also specify a complex number, it is also possible to specify a complex number by a positive real number (the distance OP) and an angle. Suppose the complex number (a, b) is specified by a distance r and an angle x and the complex number (c, d) is specified by a distance p and an angle y. It turns out that the complex number $(a, b) \times (c, d)$ is represented by the distance rp and the angle $x + y$.

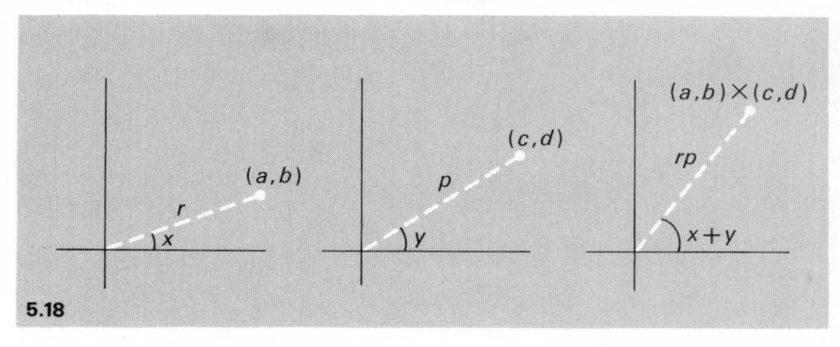

5.18

So the geometric interpretation of multiplying two complex numbers is to multiply the distances and add the angles – really quite a startingly simple result when you look at the definition in terms of pairs:

$$(a, b) \times (c, d) = (ac - bd, ad + bc).$$

We therefore have a geometric interpretation of the operations of addition and multiplication of complex numbers. Regarded as transformations of the points of the plane, *addition* corresponds to a *translation* of the points and multiplication corresponds to a *rotation* together with a *stretching*. For example, the addition of (a, b) to any complex number has the effect of translating the point it represents a units horizontally and b units vertically.

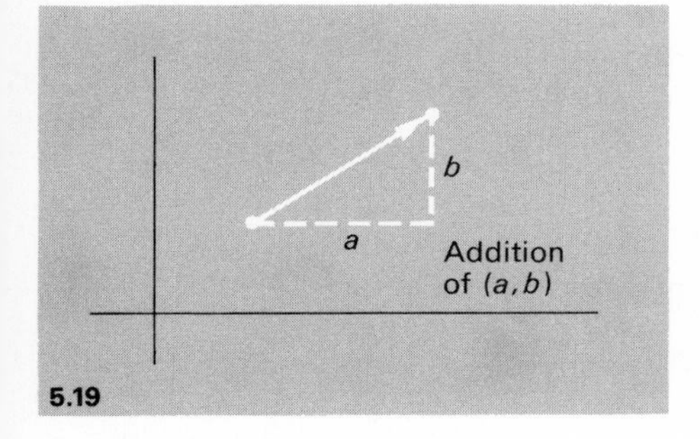

5.19

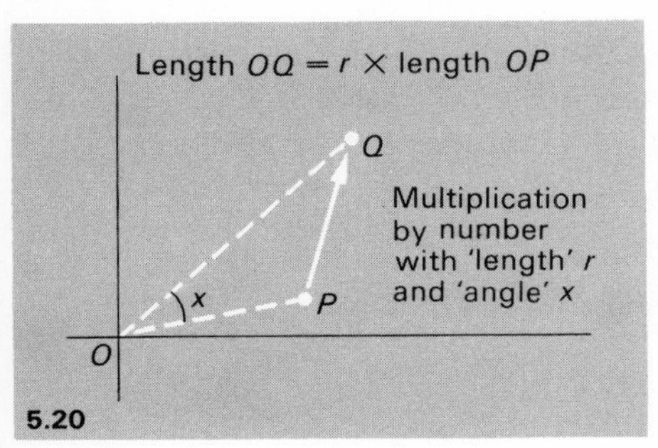

5.20

Multiplication of any complex number by the complex number specified by 'distance r, angle x' has the effect of rotating points by an angle x and stretching distances by a factor r. (If r is less than 1, then the 'stretching' becomes a 'contraction'.)

You may like to check that this interpretation is not inconsistent with a geometric interpretation of ordinary addition and multiplication of real numbers in terms of points on the number line.

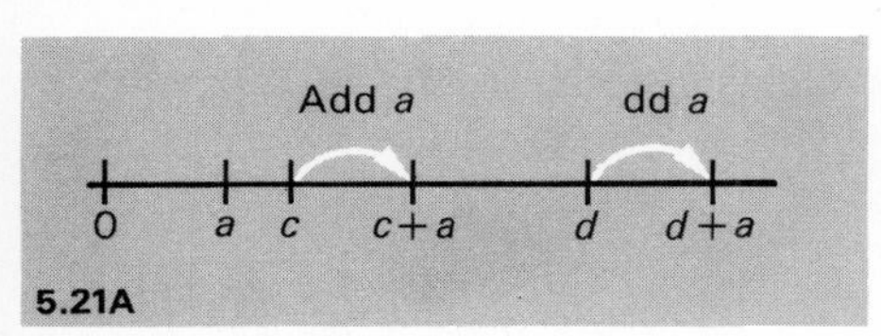

5.21A

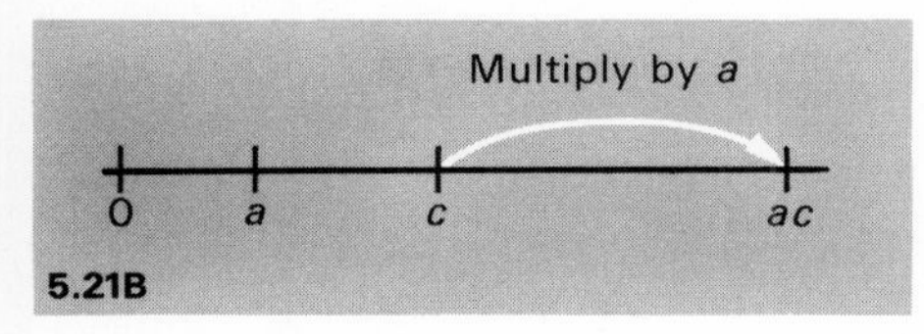

5.21B

I have several times tried to emphasize that even though the complex numbers do form a new number system, they can be thought of as an extension of the real-number system. This is so in two senses – the real numbers are 'embedded' in the complex numbers and a complex number is specified as a pair of real numbers. In a way, the complex numbers throw open a whole new dimension into understanding mathematical ideas. We have already seen a simple example of this. The picture of multiplication and addition of real numbers that we can get from the number line is just a 'collapsed' view of multiplication and addition of complex numbers that we get by interpreting a complex number as a point in a plane. If you live in the world of real numbers alone you are in a half-world. That world is difficult and exciting enough, so you can begin to imagine the fascination that accompanies each new step into the world of mathematics.

Patterns of complex numbers

We conclude this chapter by exposing a relationship between the geometric interpretation of complex numbers and some of the work we did in the previous chapter on symmetry. We begin by commenting on a remark that we made earlier. We said that every equation of the form

$$a_n x^n + a_{n-1} x^{n-1} + \ldots + a_0 = 0$$

has n solutions in the set of complex numbers. Let's take a fairly simple example: the equation

$$x^4 - 1 = 0 \text{ or, equivalently, } x^4 = 1\,.$$

It is easy to spot two solutions; $x = 1$ and $x = -1$ each satisfy the equation, because both $(1)^4$ and $(-1)^4$ are equal to 1. But what about the other two? Well, think for a moment about the geometric significance of the equation $x^4 = 1$. We are looking for a complex number which, when multiplied by itself four times produces exactly the number 1. Suppose the number has 'length' p and 'angle' q.

Remember that multiplying complex numbers multiplies their lengths and adds their angles. This means that x^4 has length p^4 and angle $4q$.

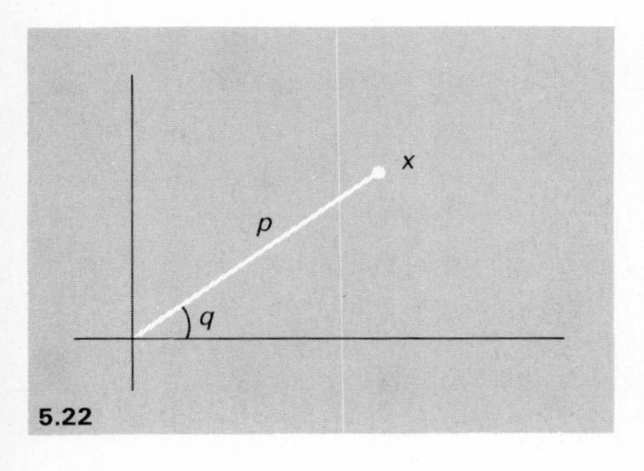

5.22

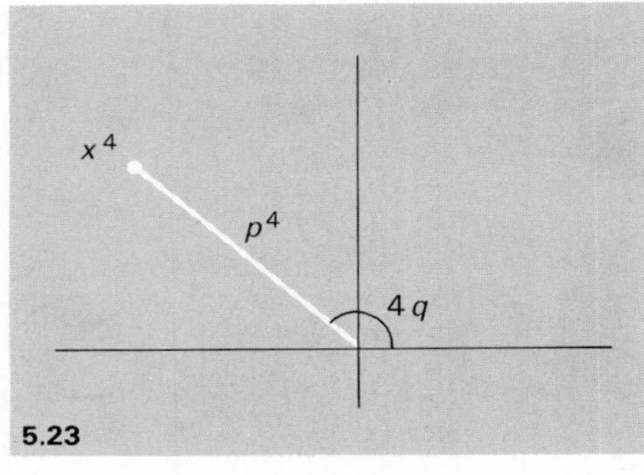

5.23

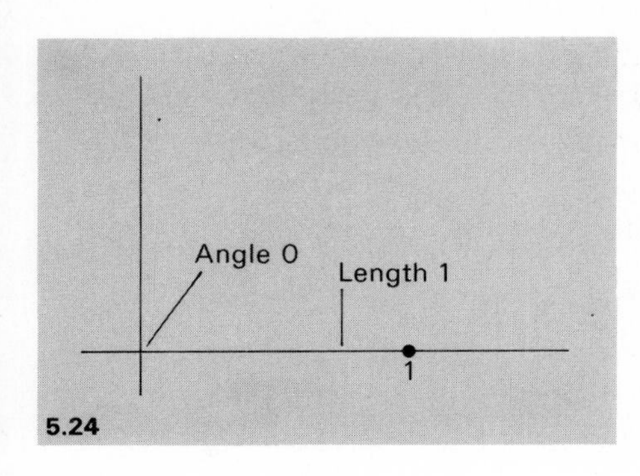

5.24

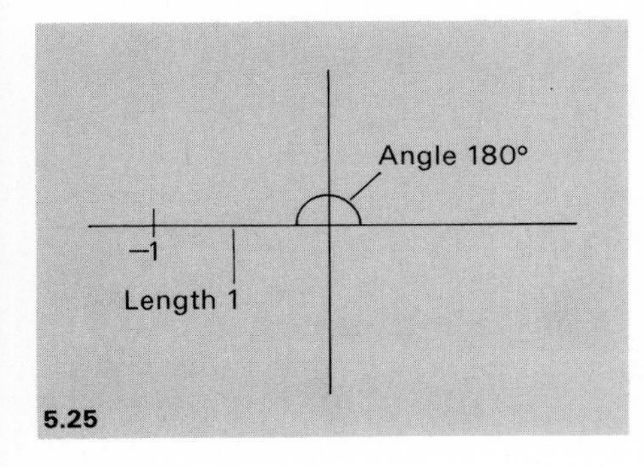

5.25

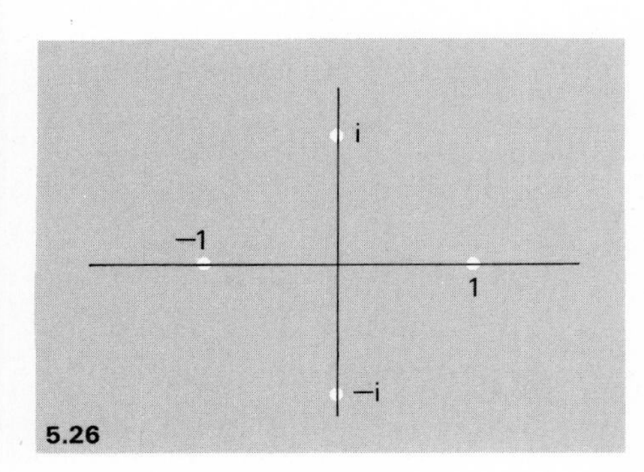

5.26

But we want x^4 to be 1 and 1 is the number with length 1 and angle 0.

The two solutions we already have fit this bill – the solution $x = 1$ is obvious. It has length 1 and angle 0, so x^4 also has length 1 and angle 0, as required. Now try $x = -1$. It has length 1 and angle 180°.

So x^4 also has length 1 and it has angle $4 \times 180°$, which is 720°. And since 720° corresponds to 2 complete revolutions of 360°; an angle of 720° is equivalent to an angle of 0. So x^4 corresponds to the point 1 as required.

We can use the same ideas to find the other two numbers for which $x^4 = 1$: we try to specify the lengths and the angles of such numbers. The length of any such number must be a positive number which, when multiplied by itself four times, must be 1. The only possibility is that the length is itself 1. What about the angle? Four times the angle must be equivalent to a multiple of 360°. The only possibilities are to choose the angles 90° and 270°. So the four solutions of $x^4 = 1$ can be plotted as in Fig. 5.26.

Notice that each of these points lies on the circle with radius 1, and centre the origin (because they all have length 1) and they are equally spaced around the circle. This is not just a coincidence: it is a particular example of a general phenomenon.

The five solutions of $x^5 = 1$, the six solutions of $x^6 = 1$, the n solutions of $x^n = 1$, all exhibit the same pattern. Each equation has $x = 1$ as one solution: the others are all equally spaced around the circle of radius 1.

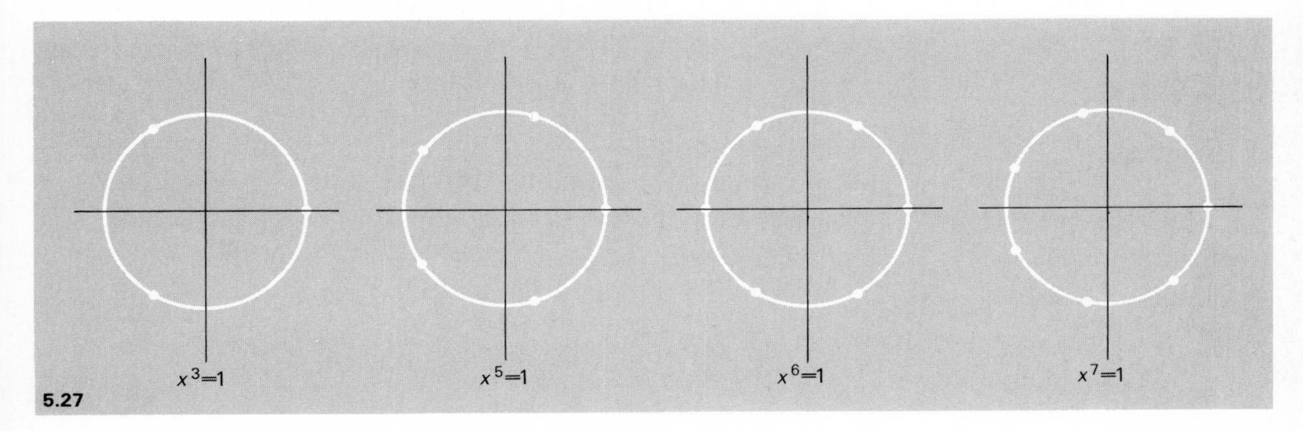

5.27

Each of these cases exhibit a certain satisfying symmetry; something that we investigated in the previous chapter. The symmetry is more evident if we join up the points.

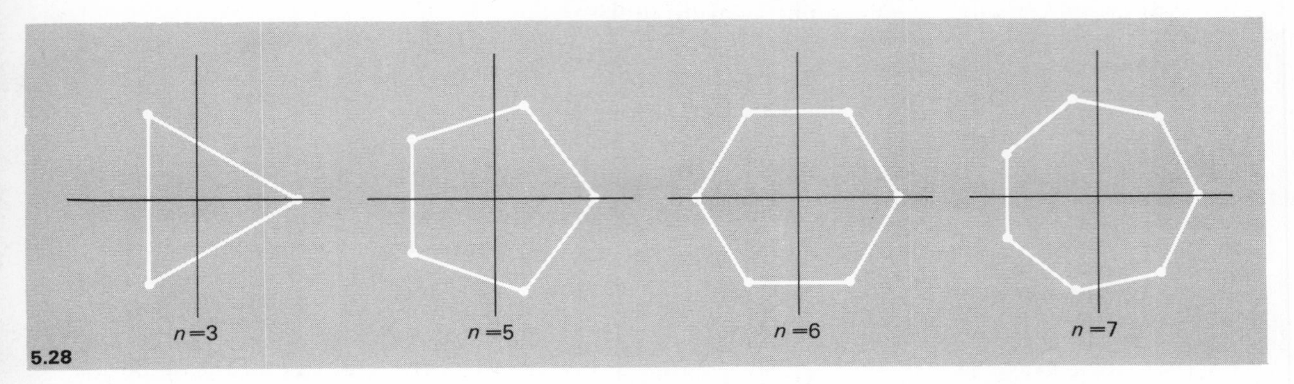

5.28

Previously we identified the symmetry of a figure with a set of functions. You may remember that we considered both reflectional symmetries and rotational symmetries. The rotational symmetries in these particular examples can be modelled in an alternative manner. We have already seen and used the fact that when a complex number is multiplied by another complex number the effect can be modelled by a stretching and a rotation.

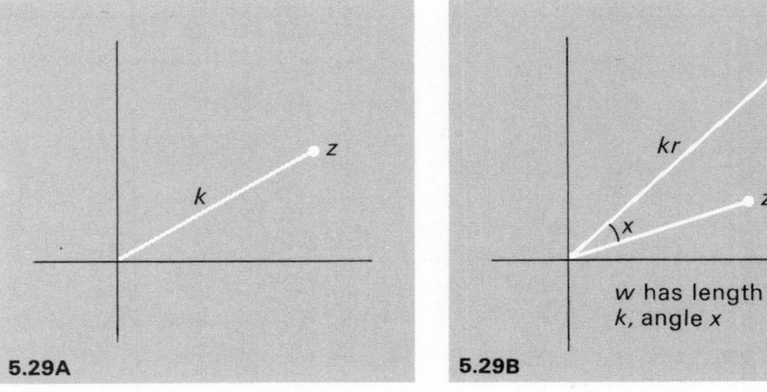

5.29A 5.29B

Let's use this idea on one of our examples – the one where $n = 5$ will do. All the numbers have length 1 and all the angles marked are the same (equal to 72°). Now think about z for a moment: since it has length 1 and angle 72° multiplying z by itself produces a rotation by 72°, with no stretching. So it takes us from z to the next point round. In other words the next number is just z^2. The next one is z^3 and so on – the last is z^5, which as we know, is equal to 1.

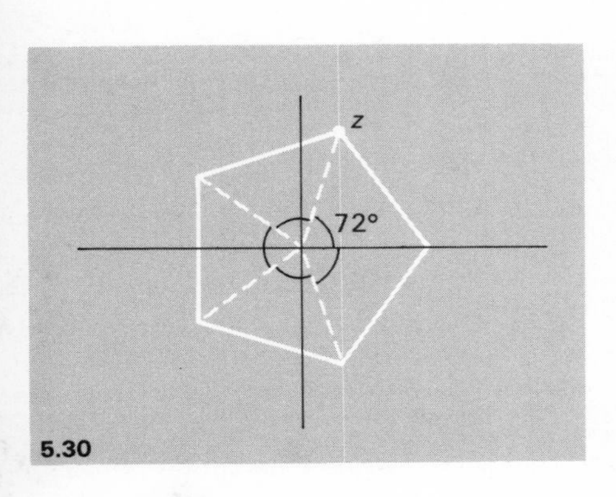

5.30

So once you know the one solution, z, to $z^5 = 1$, it can be used to generate all the others.

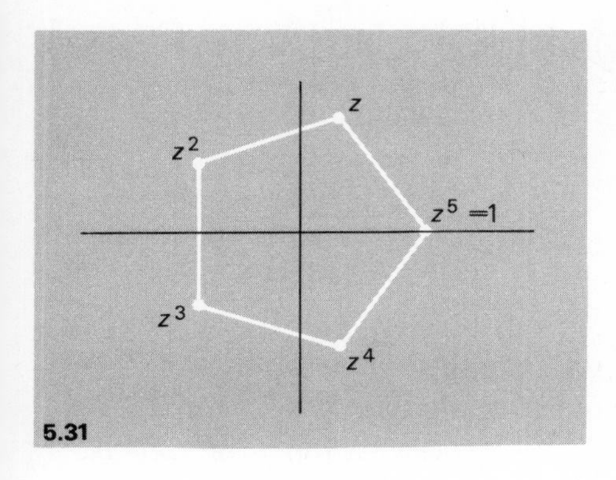

5.31

In fact, if you select any two complex numbers from the set of solutions, $(1, z, z^2, z^3, z^4\}$ and multiply them together, we get another number in the same set. We can exhibit this using a multiplication table:

	1	x	x^2	x^3	x^4
1	1	z	z^2	z^3	z^4
z	z	z^2	z^3	z^4	1
z^2	z^2	z^3	z^4	1	z
z^3	z^3	z^4	1	z	z^2
z^4	z^4	1	z	z^2	z^3

There is a marked similarity between this table and the table we saw in the previous chapter describing symmetries: that similarity is in itself a strong hint that there is much of interest to investigate in the system of complex numbers. In fact, it is just the tip of a very large iceberg. Complex numbers are one of the most useful practical tools – required reading for any physicist or engineer – and the most stimulating of intellectual foods, but we have other objectives in this present book, and so we have to satisfy ourselves with this introduction to the idea.

6 The mathematics of motion

When going on a long journey, have you ever indulged in frequent recalculations of your average speed so far for the journey? Just what do we mean by speed? Simply the *rate* at which distance is being covered. An average speed is simply the speed you would have to have been travelling if your speed for the journey had been constant, unchanging. But think a little deeper. What do I mean when I say 'travelling at a constant speed'? Presumably, the reading on the speedometer would have to be steady. So what does the reading on the speedometer mean? It is the speed you are travelling at a particular instant. But what does that mean? What is an instantaneous speed? If speed is a *rate* at which distance is covered, then the only way to measure it would appear to be to calculate the time to travel a certain distance and divide the distance by time. But this then begs the whole question of 'instantaneous' speed. If time elapses for its measurement then the 'instant' has passed. Of course a speedometer measures over such a small interval of time that to all intents and purposes it is instantaneous – just as is a police radar speed trap. All we actually measure is an average speed, but the smaller the time interval the more closely it matches what we would like to call the instantaneous speed.

When mathematics is used to tackle problems involving motion, these difficulties have to be faced up to and it is an explanation of the idea of an instantaneous rate of change

that is at the heart of the calculus. It turns out that a key to this explanation is the understanding of the difficulties that we outlined in the previous chapter over the definition of irrational numbers.

In this chapter I shall try to demonstrate some of the characteristics of the calculus. We shall meet the notion of a *limit* – the concept that is required to put both the calculus and irrational numbers on a sound footing – in the next chapter.

Sir Isaac Newton (1642–1727) was probably the greatest intellect ever produced in England. Much has been written of his incredible output, but it is difficult today to understand just how revolutionary his thoughts were, just how much he was able to break away from the conventional methods of thought of his time. The notion of gravity is familiar to us all–some of the greatest minds of the time were unable to grasp Newton's ideas.
Lord Keynes said 'Newton was not the first of the age of reason. He was the last of the magicians, the last of the Babylonions and Sumerians, the last great mind which looked out on the visible and intellectual world with the same eyes as those who began to build our intellectual inheritance rather less than 10 000 years ago'.

There can be no doubt that the calculus is one of the most important sets of techniques in mathematics. This is not the place for a systematic treatment of the subject, but you should be able to get a feel for the distinguishing features of the calculus. If you work through this chapter with some degree of understanding you can be assured that you have made a significant step; you will have crossed the threshold into the arena of mathematics that has been and will continue to be the essential requirement for the solution of many of the problems of science and technology.

The first systematic account of the methods of the calculus were produced by Newton and Leibnitz – two of the most towering intellectual minds not only of the seventeenth century but of all time. The fact that both men announced their discoveries within a short space of time prompted an unfortunate controversy, with accusations of plagiarism. But it is by no means exceptional for discoveries to be made

G. W. von Leibnitz (1646–1716) made considerable contributions to mathematics but is unfortunately often remembered most vividly for the controversey over the 'invention' of the calculus. But it is his notation for the calculus, which he introduced in 1675, which has survived.

almost simultaneously, and in any case the calculus did not come out of nowhere as a finished article. Much preliminary work had been published previously and there was considerable activity aimed at tackling problems involving motion – seeking an explanation of a variety of phenomena from the motion of the planets to the motion of a projectile. There was no mathematical technique to describe these situations, and so the work to invent one was intense.

Calculating speeds

Our reference to problems involving motion is pertinent because, although the calculus has applications in many diverse situations, those involving *change* are its *métier*. The most direct application is to consider just what is meant by the *speed* of a moving object.

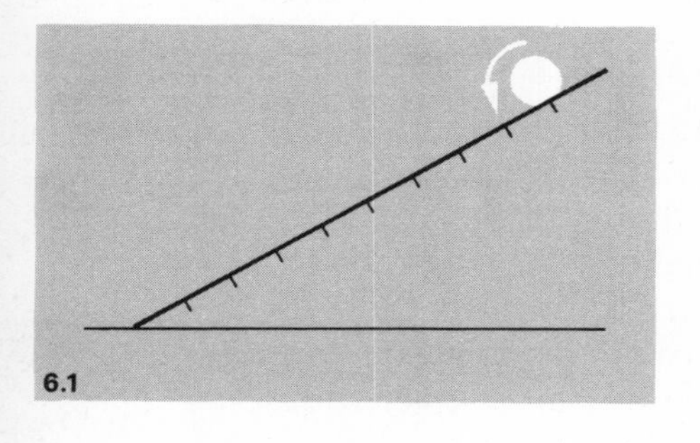
6.1

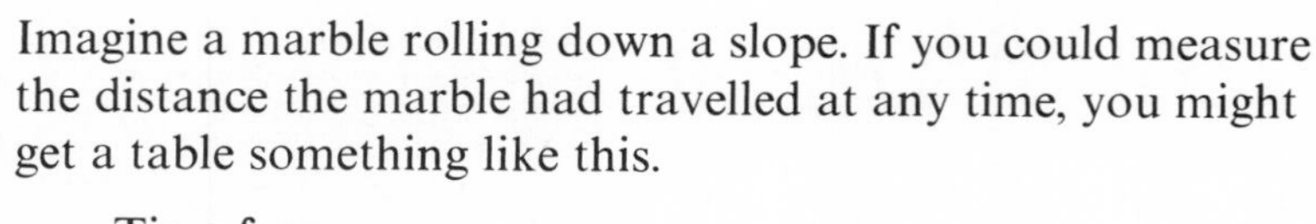

Imagine a marble rolling down a slope. If you could measure the distance the marble had travelled at any time, you might get a table something like this.

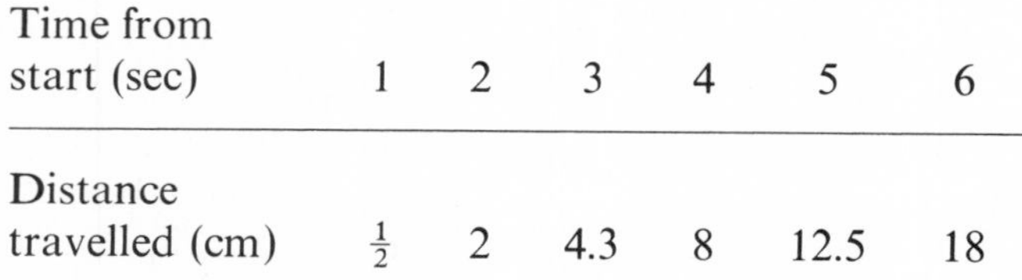

Time from start (sec)	1	2	3	4	5	6
Distance travelled (cm)	$\frac{1}{2}$	2	4.3	8	12.5	18

Such a table certainly gives some idea of the speed of the marble. The larger the distance travelled in any one-second time interval, the faster the marble must have been travelling. For example, between the second and third second it travelled 2.3 cm, but between the fourth and fifth second it travelled 4.5 cm, so it was obviously travelling faster in this time interval than in the earlier one.

We come to this conclusion by calculating, consciously or subconsciously, the *average* speed of the marble during the time intervals in question. The average speed is just the distance travelled divided by the time taken to travel the distance. But the average speed may not tell you much about the *actual* speed that the marble is travelling at any particular instant. If you measure the distance travelled over a very small interval of time, then the average speed does tell you quite a lot, because you make the assumption that variations of speed over a short time are not likely to be great. At the other extreme, if you were told the average speed of a car over a 4-hour journey, you would be extremely cautious about saying what the speed was at any particular stage of the journey.

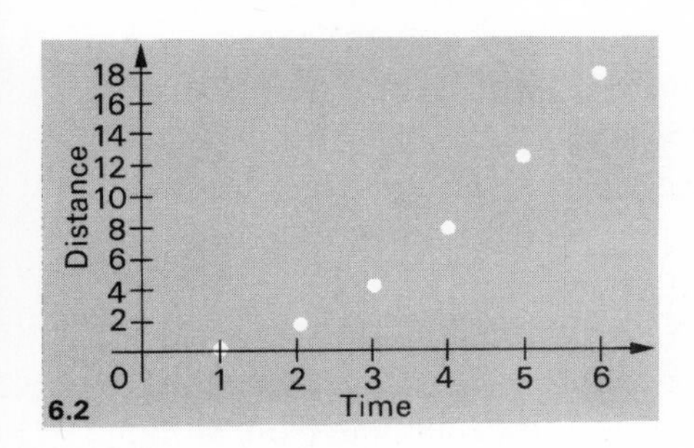

6.2

These ideas can be illustrated graphically. If we plot the data for the moving marble on a graph we get the set of points in Fig. 6.2.

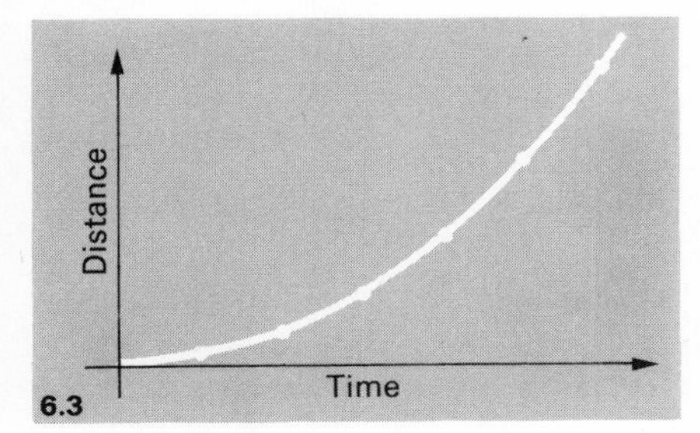

6.3

From what we know about the nature of the motion, it would seem reasonable to get a complete picture of the motion by joining up the points to give a smooth curve. The average speed for the complete motion is the total distance travelled (18 cm) divided by the total time taken (6 seconds), so it is 3 cm per second. If the marble had travelled at a steady, constant, speed of 3 cm per second then, in each one-second interval it would travel 3 cm: the total distance travelled, recorded at one-second intervals, would be as follows

Time from start (sec)	1	2	3	4	5	6
Total distance travelled (cm)	3	6	9	12	15	18

giving a set of points like this.

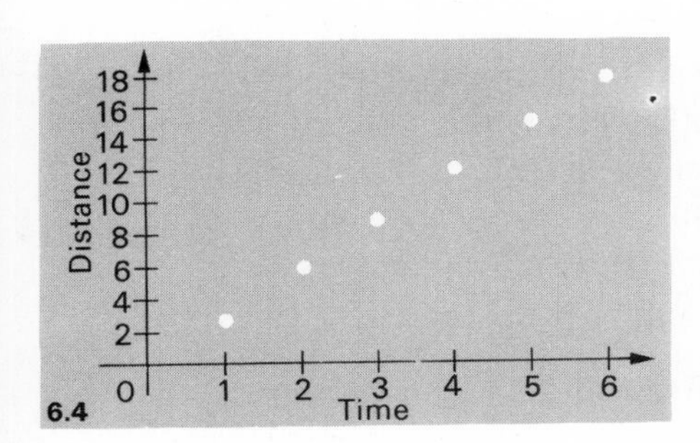

6.4

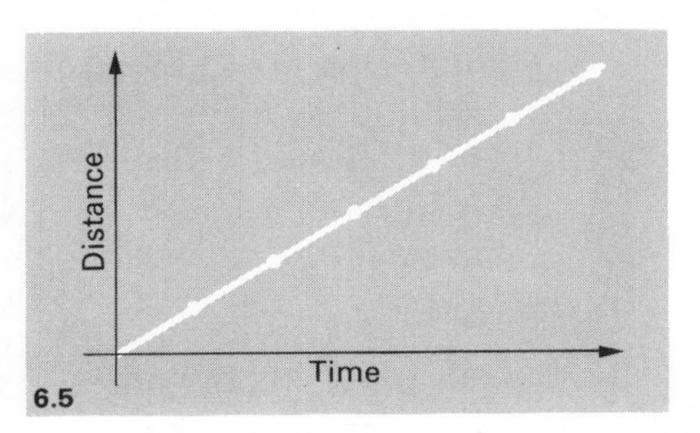

6.5

And when you join them up into a smooth curve what you get is a straight line.

A distance–time graph is a straight line for constant speed motion.

If you compare the two graphs for each of the different types of motion you can see that the steady, constant speed, graph is just the straight line joining the end points of the other graph. The fact the constant speed graph is a straight line is not an accident – it is a characteristic of this type of motion. You can illustrate the average speed over any other time interval in the same way, by joining appropriate points with a straight line. The average speed over the appropriate time interval is given by the steepness of the line, which is termed the *slope* of the straight line.

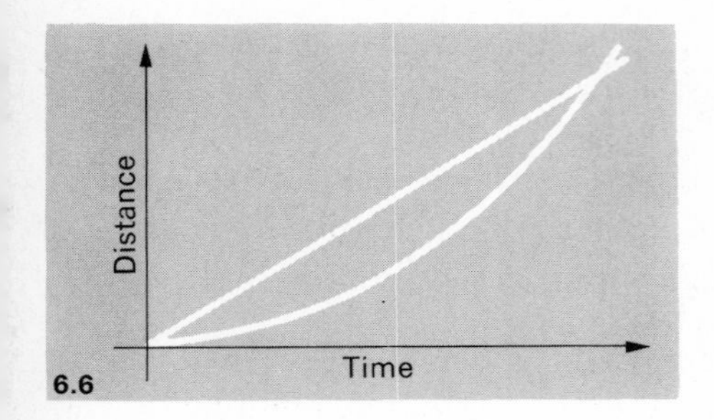

6.6

Do you see how the slope gets steeper – the ball is 'picking up speed', or accelerating as it rolls down. Remember that the curved graph describes the motion of the marble rolling down a slope.

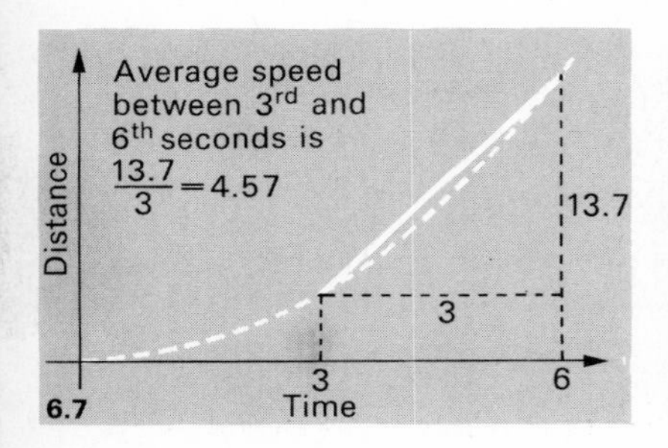

6.7

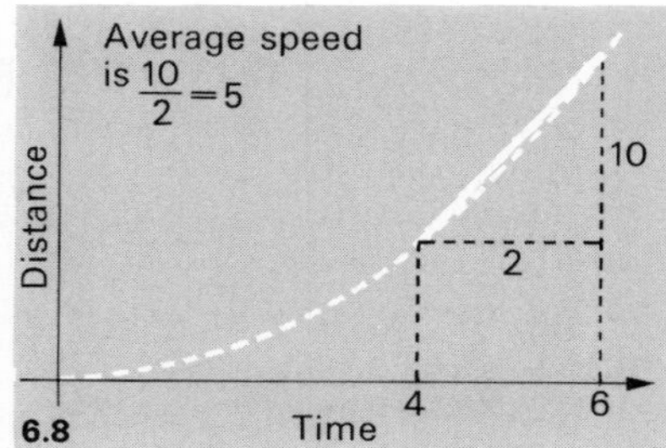

6.8

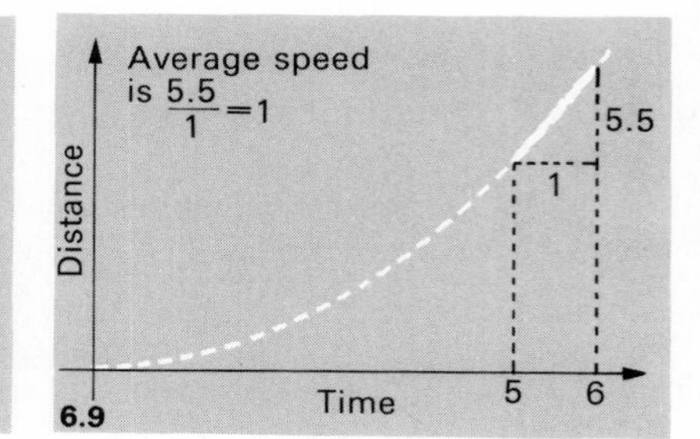

6.9

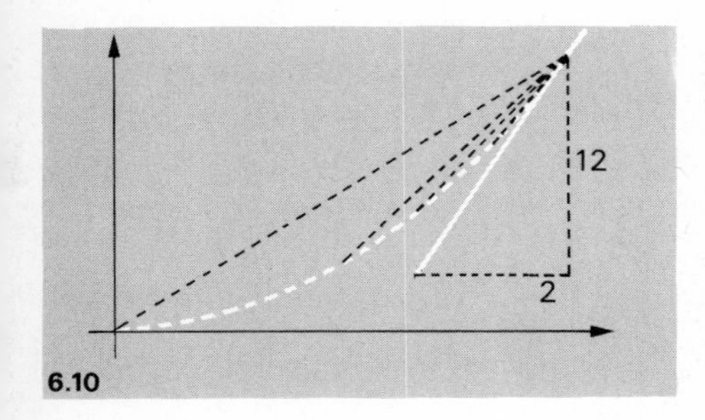

6.10

Suppose you were asked to estimate the speed at which the marble was travelling at the sixth second after starting. How would you approach that question?

What straight line are they all getting closer to? The tangent to the curve at the point in question. The gradient of this tangent is what is defined to be the instantaneous speed of the marble at the sixth second. The tangent is a visualization of what we mean by an instantaneous rate of change – an average speed calculated over a zero interval of time! The gradient can be measured on the diagram – it is $\frac{12}{2}$, which is 6.

The gradient or slope of a tangent gives a measure of the slope of the graph itself at the point where the tangent is drawn. Indeed we *define* the slope of a graph at any particular point as the slope of the tangent at that point.

Making it more precise

The disadvantage of this approach is that the answer you get is very dependent on the accuracy with which the diagram is drawn. There is little option in this particular example, because the data is rather crude anyway; all we have is a set of six measurements to go on. But suppose that we actually had a formula to describe the motion. Suppose, for example, the slope of the ramp were adjusted so that after t seconds the marble travelled exactly t^2 centimetres. (This is not such an outrageous proposition as you might think: a body which is dropped and then just falls under the force of gravity travels approximately 490 t^2 centimetres in t seconds.)

If distance travelled in t seconds is t^2 you get a table like this

Time	0	1	2	3	4	5
Distance	0	1	4	9	16	25

In a case such as this, with the motion described algebraically, by a formula, it would seem that it should be possible to find another formula which would enable you to *calculate* the slope of the graph at any point by feeding in numbers to the formula, rather than estimate it by drawing. We *can* do this and that is our next job. When you have seen how it is done you will have in your minds the essentials of one of the most powerful of all mathematical tools.

Deriving a formula

The way we approach it is to go through the same process as before, but translating the geometry into algebra. In this way we shall obtain a general procedure – one that we can use to find the slope at any point on the graph of (more or less) any function. Let's start by taking the graph of t^2 and finding the slope of the tangent at the point, P, where $t = 6$. If you work out a table of values for t and t^2 and plot the graph, you will get something like this.

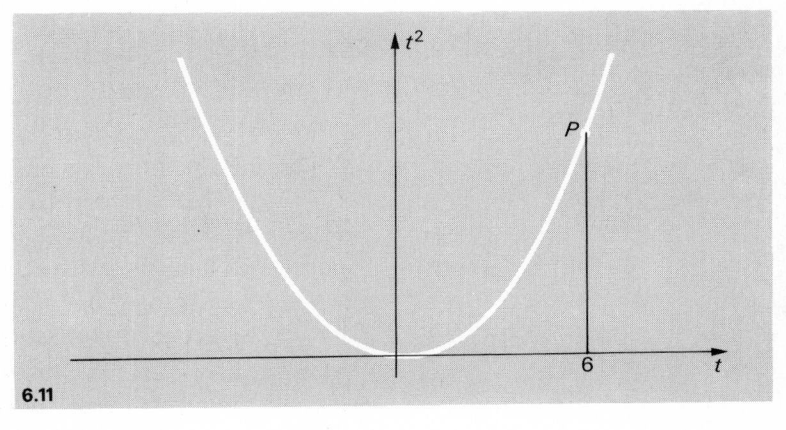

6.11

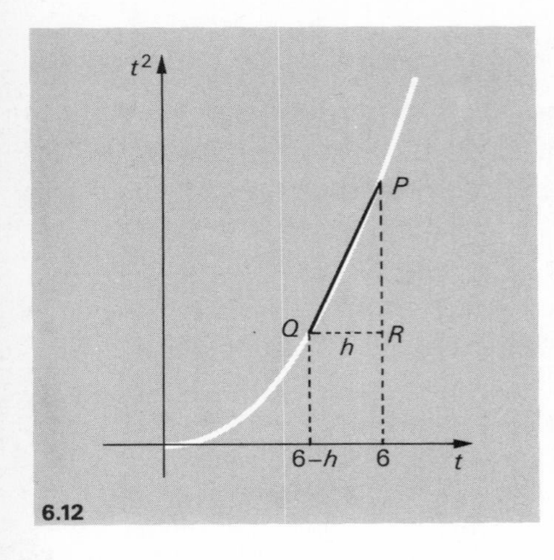

6.12

Remember that this is not just an abstract graph – it is a representation of something familiar and real – it pictures the action of a marble rolling down a ramp. Don't be confused, though, by the shape of the graph – it is *not* a literal picture of the ramp, it describes distance travelled for any value of time from starting. First, we take an estimate of the slope of the line joining the point P to another point on the graph close to it, say the point where t has the value $6-h$. Remember we are going to go through the same procedure that we went through before so as to get an algebraic description of the tangent. Call this point Q and draw a right-angled triangle PQR, as in the diagram. Then the slope of PQ is measured just by the ratio $\frac{PR}{QR}$.

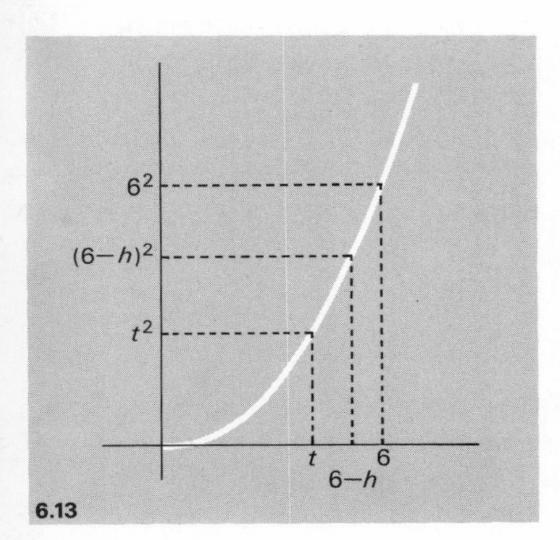

6.13

The length of QR is easy, it is just h. We can work out the length of PR by calculating the heights of P and Q above the horizontal axis. Remember that the graph is the graph of t^2, so the height of any point above the horizontal axis is given by t^2. So the height of P above the axis is 6^2, which is 36, and the height of Q above the axis is $(6-h)^2$. So PR, which is the difference between these, is $36-(6-h)^2$.

This can be simplified a little, by multiplying out the term $(6-h)^2$ – remember we worked out a formula for this sort of expression in Chapter 2. It turns out that

$$(6-h)^2 = 36-12h+h^2,$$

so $36-(6-h)^2 = 36-(36-12h+h^2)$

which is $12h-h^2$.

On p. 42 we had the formula $(x+y)^2 = x^2+2xy+y^2$

Now we can calculate $\frac{PR}{QR}$: it is

$$\frac{12h-h^2}{h}, \text{ which is } 12-h.$$

Think now what we have done, we have worked out the slope of a line joining P to another point on the curve near to P. But this other point Q, was not a particular point on the curve – it was simply the point corresponding to giving t the value $6-h$. By varying h, the point Q can be made to move along the curve. In particular, by making h smaller, we can make Q get closer to P; and this is precisely the action that we want to model with our mathematics. The use of the symbol h means that we can now analyse that behaviour algebraically rather than geometrically.

Remember that we want to calculate the slope of the tangent to the curve at the point P and that the closer Q gets to P, so the more nearly does the line through Q and P approach to that tangent. But look at the expression we got for the slope of QP: it was

$$12-h.$$

What happens to this when h gets small? It gets closer and closer to the number 12. So this number 12, must be the value of the slope of the tangent – there is no other possibility.

If, as we proposed earlier, this graph described the distance travelled by a marble rolling down a slope, then what we have been able to do is to *calculate* the instantaneous speed at which the marble was travelling at the sixth second.

That is quite an achievement, but there is more to it than that. The process we went through is completely general, we could have applied it to any other point on the graph *and* to any other graph (well, almost; some graphs give problems which we shall mention later). This second remark is particularly interesting because it means that whenever we have a *formula* which links the distance travelled to the time of travel then we can *calculate* the speed at any instant. And such formulae can often be obtained by theoretical arguments, applying physical principles. For example, it is perfectly feasible to work out a formula for the distance travelled in any particular time by a cricket ball thrown at a specified angle with a specified speed. From this you could calculate its speed at any point in its trajectory.

Obviously, there are many more important practical cases where the technique is vital – calculating the speed of a spacecraft, or a moving part on a piece of machinery, and so on. (In these cases, you very often *only* have the formula so measurements are not available – so a theoretical analysis is, at least on the face of it, the most appropriate approach. And in a minute, we shall see that it is not just speeds that the technique applies to.)

Let's just take stock of where we have got to. The basic problem we have been worrying over is that of getting to grips with the notion of instantaneous speed. The difficulty of trying to 'measure' over a zero time interval is reflected by the graphical interpretation. Here, average speed corresponds to the slope of a line which cuts the curve in two places, whereas instantaneous speed corresponds to the slope of a tangent – which touches the curve at only one point. That is a satisfactory explanation if we are content to draw curves and tangents but it can only ever give an approximate answer, because of the inaccuracies inherent in any drawing. To seek a precise description we turn to an algebraic description, where the curve is specified precisely, by a formula. What we then seek is a formula for the slope of the tangent, and we get that by going through the geometric process again, but expressing it in algebraic terms. But this still gave us the answer only for one particular point on the graph. Nevertheless, it is the breakthrough, because we can now generalize the technique so that we can apply it to any graph.

Generalizing the argument

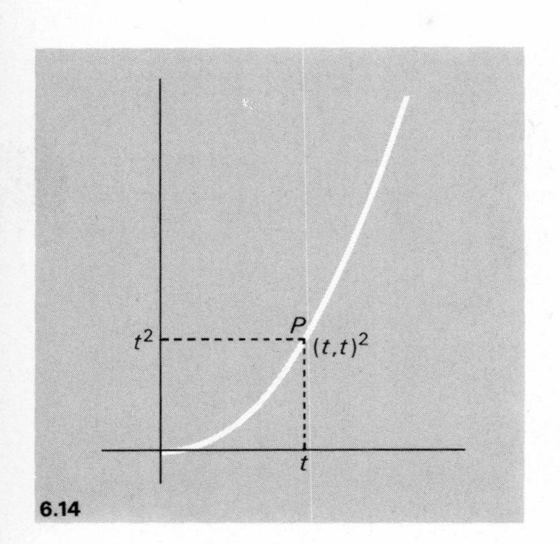

6.14

The easiest way to generalize the argument is to take it in two stages: first we shall stick with the same graph as before but find the slope of the tangent at an arbitrary point. Then we shall extend the technique to more general graphs. In both cases we have only to go through the same piece of algebra that we have done for the t^2 graph and the point where $t = 6$ but replacing specific numbers by more general symbols.

First, then, we shall calculate the slope of the tangent at an arbitrary point on the graph of t^2. Such an arbitrary point has coordinates – 'map references' – (t, t^2).

Before we start, let's just clear up a small point. Previously we took, as a value of t, close to 6, the value $6-h$. This gave us a point to the left of P. The reason for this was that it tied in with the way we were discussing the problem on p. 158 in the numerical, approximate, example. But there is no reason for taking a point to the left of P: it could be to the left or to the right. To allow for this, it is usual to take as a value close to t the quantity $t+h$ where the understanding is that h can be positive or negative – giving points to the right or to the left of P.

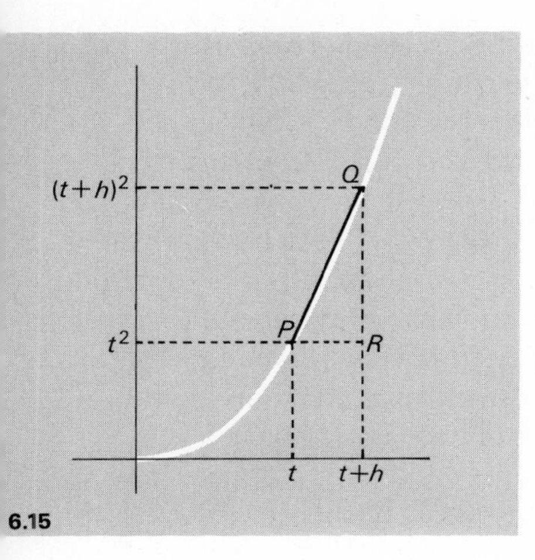

6.15

Again we use the formula $(x+y)^2 = x^2 + xy + y^2$

So let's take the point Q, close to P, as the point corresponding to the value $t+h$: this is the point with coordinates $(t+h)$ and $(t+h)^2$.

	Special case	General case
Point P	$(6, 6^2)$	(t, t^2)
Point Q	$(6-h, (6-h)^2)$	$(t+h, (t+h)^2)$

Now we have to work out the slope of the line through P and Q, which is the ratio $\frac{QR}{PR}$. As before, PR is just h, and QR is the difference between the heights of P and Q above the horizontal axis. So

$$QR = (t+h)^2 - t^2.$$

We can work out $(t+h)^2$ using our formula, just as we did for $(6-h)^2$ before. It comes to $t^2 + 2ht + h^2$.

So $QR = t^2 + 2ht + h^2 - t^2$

which is $2ht + h^2$.

So the ratio $\frac{QR}{PR}$ is $\frac{2ht+h^2}{h}$ which comes to $2t+h$.

Previously, when t had the value 6, this came to $12+h$, so we have a useful check, because we get the same answer if we substitute 6 in place of t in this present result.

This expression, remember, gives the slope of the line joining the points P and Q and this line gets closer to the tangent at P as h gets smaller and smaller. The value of the slope of the actual tangent is obtained by putting h equal to zero; it is just $2t$.

The power of our new results is that it gives a formula for the slope at *any* point or, in the context of our practical example, the value of $2t$ is the value of the *speed* at *any* time t seconds after the start of the motion. And, even though more difficult practical problems have to take account of difficulties like friction and so on, the same basic technique still applies.

Any point, any curve

Now we take a major step, one that will set us well on the road to understanding the calculus. We move on to tackle the problem of finding the slope of the tangent at any point on any curve. And why is finding the slope of the tangent at any point on a curve important to us? Because it gives us the *rate of change* of whatever it is we are measuring. If it happens to be distance travelled, then the rate of change is the speed. It could be a rate of inflation in a discussion of economics, it could be a rate of increase of the population, it could be the rate of growth of a cow being fattened up, the rate of increase of pressure on a diver submerging under the sea. As usual when making generalizations we first translate the familiar and particular into a form that makes sense in a wider context and then interpret it in that context.

A f B

Remember what a function is: see p. 105.

What is the general language that we use to describe graphs, relationships and so on? The abstract concept is that of *function*, and what we are now going to do is to set out to find the slope of the tangent at any point on the graph of (almost) any function. First, let's dispose of the cautionary word 'almost' so that we can get on with the job. The point is that for some graphs it is not possible to find a tangent at every point – in particular if the graph has sharp corners it is difficult to know what is meant by the tangent at one of the 'awkward' points.

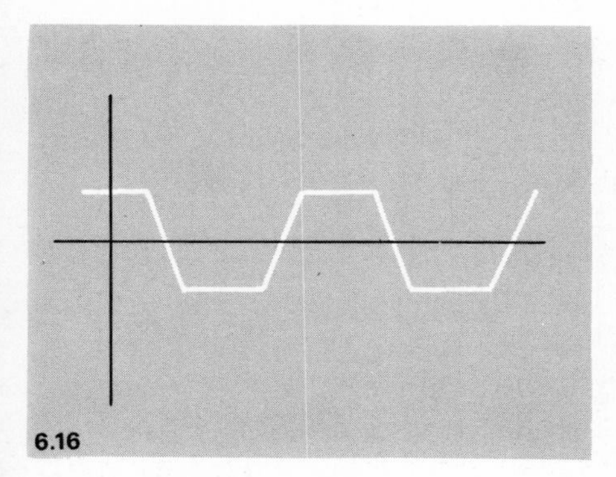
6.16

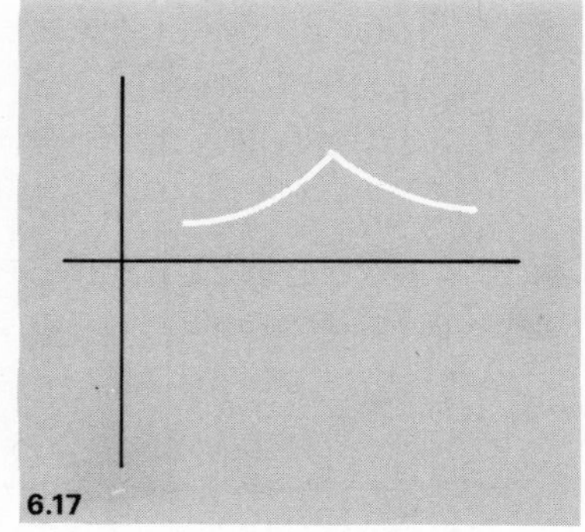
6.17

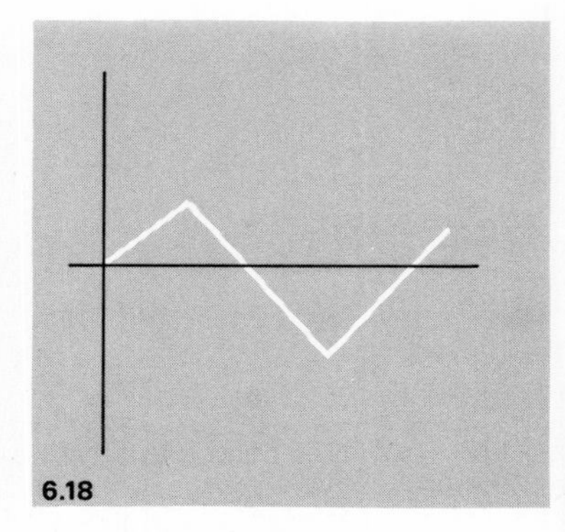
6.18

All we need to do is to rule out functions that give rise to difficulty – we are still left with a vast array of possibilities.

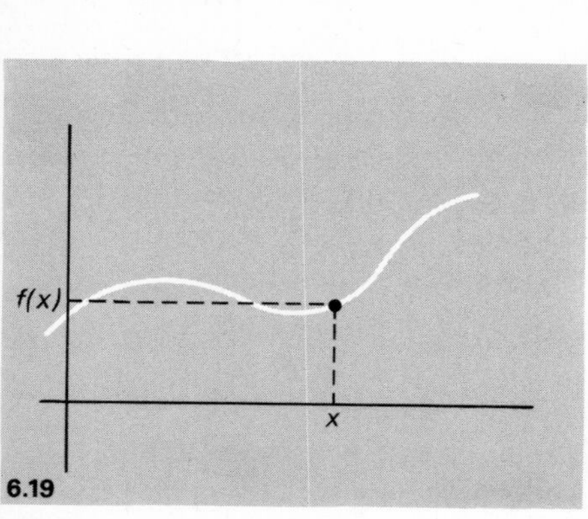

6.19

So let's suppose that we have a function, f, whose graph does not have any 'sharp corners', and let's try to find the slope of the tangent to its graph at an arbitrary point $(x, f(x))$.

We shall go through precisely the same procedure as we have been through twice – but now using a general point *and* a general graph. Again we take a point on the graph close to the point in question – the point with coordinates $(x+h, f(x+h))$.

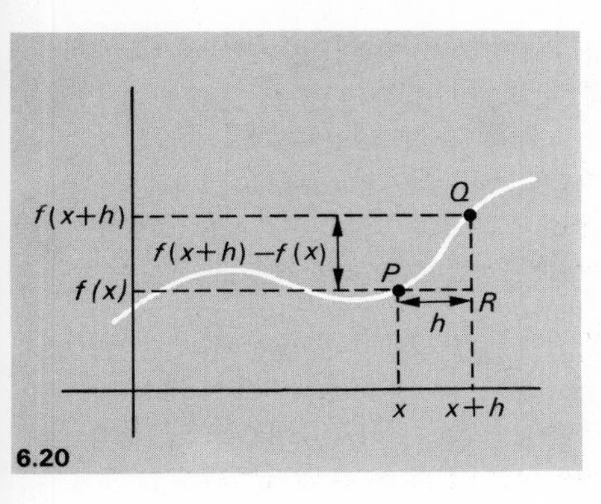

6.20

Now we write down the slope of the line through P and Q: it is $\frac{QR}{PR}$ which is $\frac{f(x+h)-f(x)}{h}$.

As before, we think of this as an approximation to the slope of the tangent at P – an approximation that is improved by making h smaller.

In each of the cases that we have considered so far, we were able to spot a special quantity to which it was clear that the approximations were approaching more and more closely. Remember in the first case, the approximation was $12+h$, in the second it was $2t+h$. It was quite clear that as h gets smaller then in the first case the approximations get closer to 12 and in the second case they approach $2t$. Quantities such as these are given a special name: we say that 12 is the *limit* of $12+h$ as h approaches 0 and $2t$ is the *limit* of $2t+h$ as h approaches 0.

Perhaps you find the phrase 'as h approaches 0' a little strange, perhaps cautious. Why can't we just 'put h equal to 0'? Well, we have to bear in mind the problem we face. We are tackling the concept of instantaneous speed, or more generally of finding an algebraic description of the slope of a tangent, and we have seen that we can only do this by creeping up on the answer by approximating to what we seek by calculating an average speed over intervals of ever decreasing size. This is the flavour that is reflected in the phrase 'approaches 0'. There is also a rather deeper point, which we shall take up in the next chapter but you can get an idea of what it is by looking at the expression $\frac{f(x+h)-f(x)}{h}$. It just does not make sense to put h equal to zero in this expression, because it is not possible to attach a meaning to division by zero. (Think about it.) Just as we have seen that 'instantaneous speed' is really an intellectual concept so is the idea of 'limit' a purely mathematical notion. Hence the rather oblique term 'approaches zero'. In the next chapter we shall see more precisely what we do mean by 'limit', but it was essentially this sort of problem – of what can actually be meant by $\frac{f(x+h)-f(x)}{h}$ when h gets closer and closer to zero – that was the substance of the objections raised against Newton's exposition of the calculus.

The intuitive concept of 'instantaneous rate of change' corresponds to the mathematical concept limit.

Because of the level of generality that we are now working at, we cannot specify the value of the limit of our approximations because the limit will vary according to the function we are dealing with: that is one of the reasons for introducing the term 'limit' – it enables us to encapsulate a general concept, applicable in many different cases, in a single word. The slope of the tangent is thus the limit of $\frac{f(x+h)-f(x)}{h}$ as h approaches zero.

But if we can be content to wait until the next chapter to formalize this notion of limit, then we can press on now to familiarize ourselves with the new idea.

Rates of change – defining a new function

The first important comment to make is that we have generalized our mathematical description of speed to a point where it is susceptible to a far more abstract interpretation which can be applied to many other situations involving a rate of change. The expression

$$\frac{f(x+h)-f(x)}{h}$$

can be calculated for any function f which is specified by a formula, whether or not it has a straightforward physical interpretation. But how do we interpret it? Well $f(x+h)-f(x)$ is the change in the value of $f(x)$ when x is increased to $x+h$ and so $\frac{f(x+h)-f(x)}{h}$ is the average rate of change over the interval x to $x+h$, just as a similar expression previously gave us average speed. And just as the limit as h approaches zero gave us the *speed*, so the limit of the average rate of change gives us the *rate of change* of $f(x)$ at the point x.

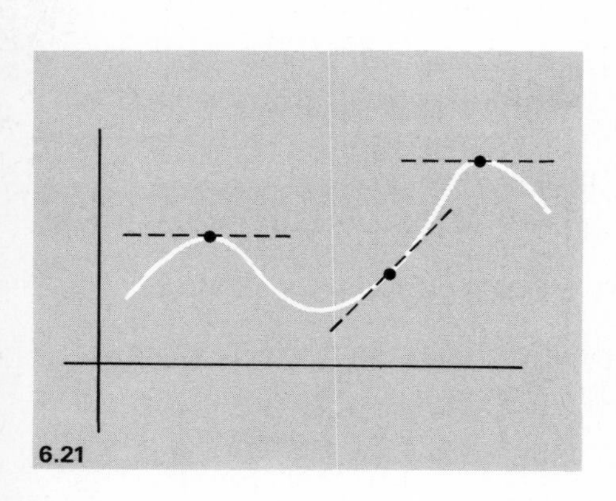
6.21

An alternative name for the rate of change of $f(x)$ at x is the *derivative* of f at x. Don't worry too much about this – the word derivative does not have its everyday meaning in this context and anyway we'll take it up in the next chapter – but there is just one other point to pick up before we go on. For any given function f the rate of change of $f(x)$ at x will depend on the value of x – choose a different point on the graph and you are likely to get a different rate of change.

If you think about it, you will see that we have in effect defined a new function, one that relates the *slope of the tangent* at the point $(x, f(x))$ to the *value of* x. For example, if f relates distance to time the new function associates speed (the slope of the tangent) with time. This new function, the *derivative* of f, is often denoted by f'. So just as

f relates x to $f(x)$

so f' relates x to the rate of change of $f(x)$ at x

that is to say $f'(x) =$ the rate of change of $f(x)$ at x.

The *derivative* enables you to work out *rates of change.*

As I have implied earlier, the notion of *rate of change* is applicable to many situations other than those concerned with motion, it can be applied to any problem concerning growth – growth of population, expansion of a gas being heated, increase or decrease of profits in a commercial concern, problems concerning inflation in economies, all involve the idea of rate of change.

In a systematic development of the calculus, the next step is to develop a set of techniques and special results for working out derivatives. An important part of this is to construct a table of derivatives for particularly useful and frequently occurring functions. We are not going into that sort of treatment here, but we will be needing a few results later, so perhaps you can accept the following list on trust.

Function (describing something that is changing)	Derivative (describing the rate) at which it is changing)
$f(x)$	$f'(x)$
k, any constant	0
x	1
x^2	$2x$
x^3	$3x^2$
x^4	$4x^3$
$\frac{1}{x}$	$-\frac{1}{x^2}$
$\frac{1}{x^2}$	$-\frac{2}{x^3}$

Solving the publisher's problem

But there is a less obvious type of application of these ideas. As an example, we can take the problem that we started way back in Chapter 3. On p. 72 we were discussing a publisher's strategy concerning the size of print area on page. You may like to flip back to refresh your memory. We had to find the maximum value of the expression $(x-6)\left(\frac{300}{x}-4\right)$ for values of x varying between 6 and 75. To help see what was going on, we sketched the graph of the function

$$f:x \to (x-6)\left(\frac{300}{x}-4\right).$$

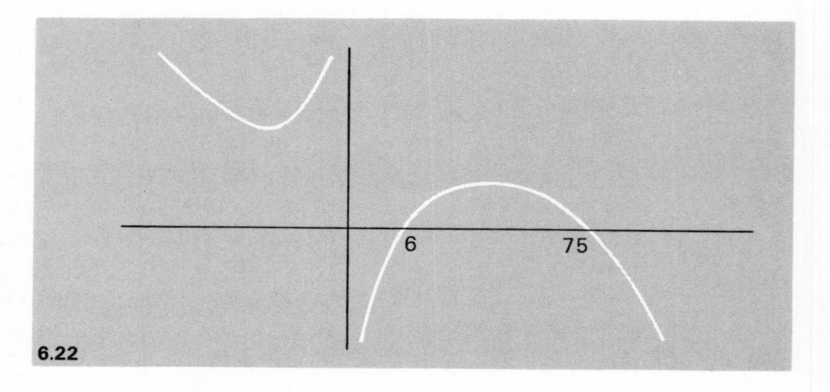

6.22

The maximum value of the expression arises when x takes the value corresponding to the point P. We now have a technique for actually calculating this value of x, because there is something particularly special about the tangent to the graph at the point P: it is horizontal. In other words, its slope is zero.

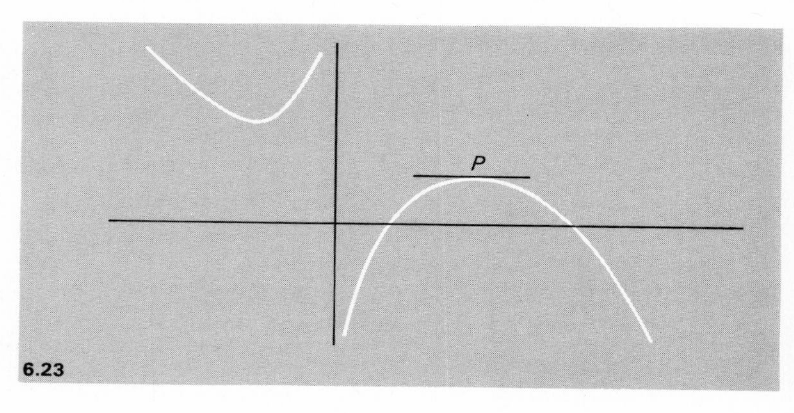

6.23

But since we know the equation of the curve, we can now work out the gradient of the tangent at any point. We know that at any point on the curve,

$$y = (x-6)\left(\frac{300}{x} - 4\right).$$

If you do some work on multiplying out the brackets you should end up with the alternative form

$$y = 324 - 4x - \frac{1800}{x}.$$

Each term on the right-hand side is an expression that we know how to differentiate (this is the term used to refer to the process of working out a derivative). So if we write

$$f(x) = 324 - 4x - \frac{1800}{x}$$

we can work out $f'(x)$: we get

$$f'(x) = -4 + \frac{1800}{x^2}$$

just by using the results in the table on p. 167.

Now we can find the value of x at the point P, because at that point we must have $f'(x) = 0$, since we know that the gradient there is zero. So at the point P, we must have

$$-4 + \frac{1800}{x^2} = 0.$$

In other words, multiplying by x^2,

$$-4x^2 + 1800 = 0$$

so $4x^2 = 1800$

which means $x^2 = 450$

so x must be about 21.2.

If you now look back again to p. 72 in chapter 3, you will be able to work out the answer to the publisher's problem – to calculate the optimum size for the area of print.

Calculating maximum and minimum values

The idea we used to solve this problem is a general one in the sense that it is applicable to a whole class of problems. If you have a function that describes the variation of a quantity as some parameter changes then if that quantity increases to a maximum value and then decreases, then the value of the parameter at that critical point can be calculated by working out the derivative of the function.

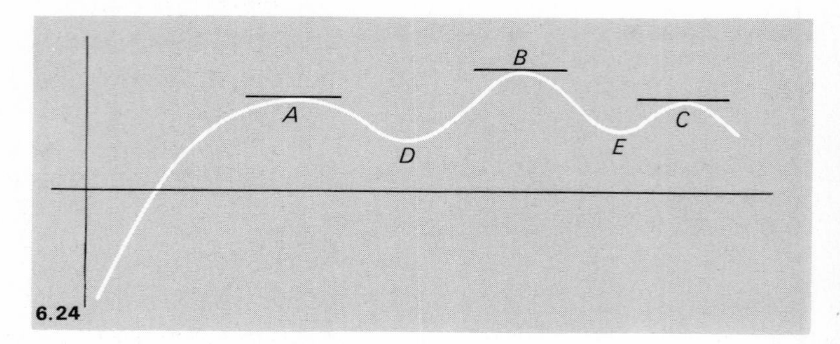

6.24

At each of the points A, B, and C the value of the derivative is zero and at each of these points the value of $f(x)$ has risen to a maximum value – not an overall maximum, but a 'local' maximum. Imagine the graph to be a graph of speed of a car plotted against time. Then the part of the graph to the left of A represents a situation in which the car is accelerating until, at A it reaches a maximum speed and starts to decelerate. At D it starts to accelerate again until it reaches another 'local' maximum speed. And so on.

There is something interesting about the points D and E also. At these points the tangents are also horizontal – the gradient of the graph is zero – and so $f'(x)$ is zero at these points. And of course these points have a similar physical significance to A, B, and C. They are again 'critical' points in the sense that there is a definite change in the character of the motion: D is the point at which the deceleration phase – from A to D – changes to an acceleration phase – from D to B.

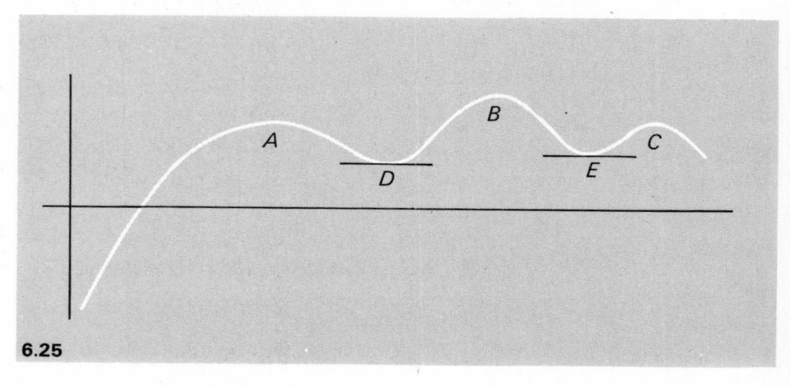

6.25

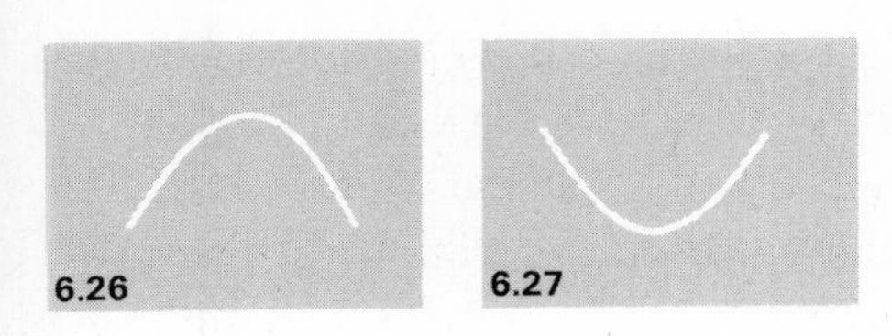
6.26 6.27

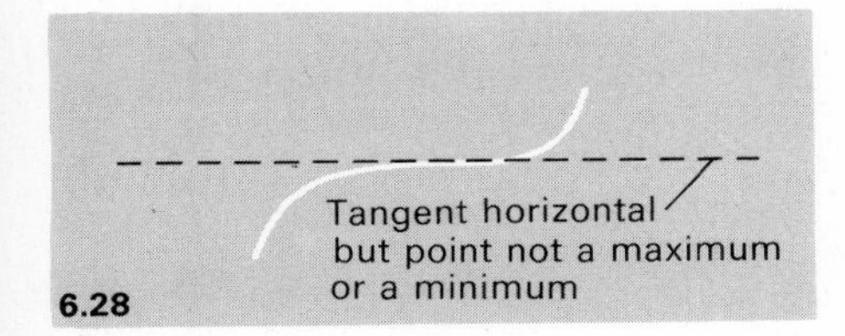

6.28

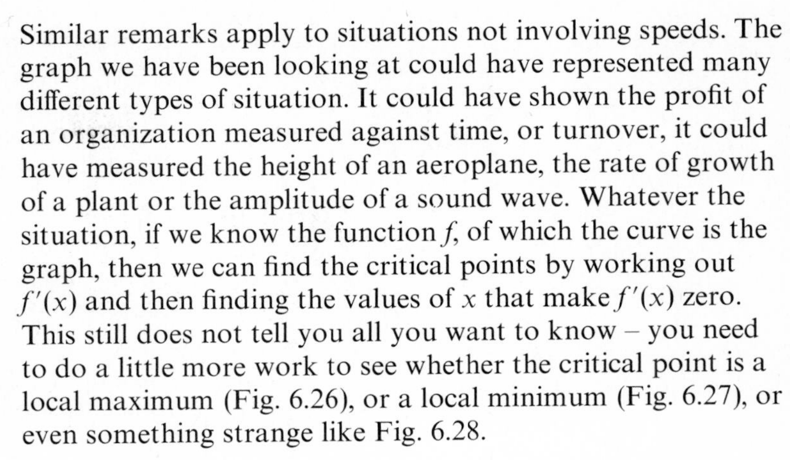

Similar remarks apply to situations not involving speeds. The graph we have been looking at could have represented many different types of situation. It could have shown the profit of an organization measured against time, or turnover, it could have measured the height of an aeroplane, the rate of growth of a plant or the amplitude of a sound wave. Whatever the situation, if we know the function f, of which the curve is the graph, then we can find the critical points by working out $f'(x)$ and then finding the values of x that make $f'(x)$ zero. This still does not tell you all you want to know – you need to do a little more work to see whether the critical point is a local maximum (Fig. 6.26), or a local minimum (Fig. 6.27), or even something strange like Fig. 6.28.

There are many many books available to help the reader to develop techniques for solving problems like this and to take the ideas further; this is not really the place to go into specialized methods. But before we go on, let's have a look at just one more problem, to give you an idea of the range of the technique.

Making tin cans

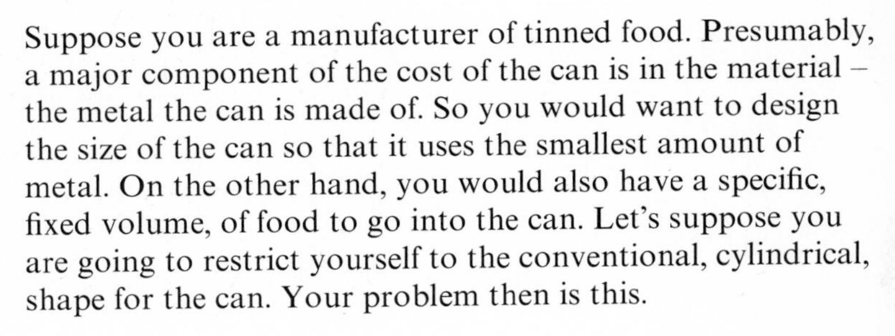

Suppose you are a manufacturer of tinned food. Presumably, a major component of the cost of the can is in the material – the metal the can is made of. So you would want to design the size of the can so that it uses the smallest amount of metal. On the other hand, you would also have a specific, fixed volume, of food to go into the can. Let's suppose you are going to restrict yourself to the conventional, cylindrical, shape for the can. Your problem then is this.

How do you choose the radius r of the can and the height h of the can so that the surface area, call it S, of the can is a minimum (so that the cost of the material is minimized), subject to the volume of the can, call it V, being fixed?

V fixed. Choose r and h to make S as small as possible.

The first step is to write down what we know about V and S in terms of r and h. The volume, V is given by

$$V = \pi r^2 h$$

and the surface area is given by

$$S = \underbrace{2\pi r^2}_{\text{(the area of the ends)}} + \underbrace{2\pi r h}_{\text{(the area of the curved surface)}}.$$

From what we have said earlier, you might now set about thinking of S as the value of some function that we can then differentiate and so apply the principles we have been discussing to find its minimum value. But there is an extra complication in this case because we know how to deal only with functions whose value depends on a single variable – as with the functions we have seen which describe the variation of speed with the variation of the single variable time. In this particular case, we have two variables r and h, contributing to the variation of S.

We can overcome this snag by using an extra piece of information that is available. We know that the volume V is fixed. In other words the quantity $\pi r^2 h$ is fixed. This means that r and h must be related – they cannot be allowed to vary independently. In particular, if the volume is fixed then if h is increased then r must be decreased. The formula for V,

$$V = \pi r^2 h$$

quantifies this observation, because we can rearrange it in the form

$$h = \frac{V}{\pi r^2}$$

which tells you exactly how to calculate h if the value of r is changed.

What is more, if we replace h by this new expression, $\frac{V}{\pi r^2}$, in the formula for S,

$$S = 2\pi r^2 + 2\pi r h,$$

we get

$$S = 2\pi r^2 + 2\pi r \times \frac{V}{\pi r^2}$$

which can be written

$$S = 2\pi r^2 + 2\frac{V}{r}.$$

Now we are in a much better position. Because V is fixed, there is only one parameter on the right-hand side that can vary – that is r. To emphasize this dependence of S on r, we can write $S = f(r)$, so that f is the function that describes the relationship between r and S. Our problem now reduces to a purely abstract one, of a type that we know how to solve. We have to find the value of r for which $f(r)$ is a minimum, where

$$f(r) = 2\pi r^2 + 2\frac{V}{r}.$$

As we know, we can tackle this by differentiating f, giving

$$f'(r) = 4\pi r - 2\frac{V}{r^2}$$

(using the table on p. 167).

We now look for values of r that make $f'(r)$ zero. These must satisfy the equation

$$4\pi r - 2\frac{V}{r^2} = 0,$$

which is equivalent to

$$4\pi r^3 - 2V = 0$$

So the value of r that we seek is that value for which

$$r^3 = \frac{V}{2\pi}.$$

So if we know what the volume, V, is, we can proceed to calculate r. But there is a little more that we can say. Remember that

$$V = \pi r^2 h.$$

If we use this expression for V in the expression we have just obtained for r^3, we get

$$r^3 = \frac{\pi r^2 h}{2\pi}$$

which amounts to $r^3 = \frac{r^2 h}{2}$.

If we divide each side of this relation by r^2 we get $r = \frac{h}{2}$.

This means that we have now arrived at a solution to the problem that does not depend on V – so whatever volume the manufacturer requires the can to be, the most economical way to make it is to have the radius equal to half the height. In other words the diameter of the tin should be the same as its height.

The fact that not many cans of food seem to have this 'square' shape is possibly a reflection on the fact that advertising psychology is a stronger influence on economics than the desire to conserve raw material.

Hopefully, you have begun to see the potential of the methods of the differential calculus. In the next chapter we shall do two things. First we shall establish the idea of limit, the cornerstone of the calculus, more firmly, and then we shall use it to develop the other strand of the subject, the one that complements differentiation – the idea of integration.

7 Going to the limit

Differentiation gives rate of change. Integration measures accumulations.

In this final chapter, we shall discuss the notion of 'limit' – an idea that has been just below the surface of a lot of the work we have been doing in earlier chapters. We have already mentioned the importance of the idea, but we shall emphasize this even more in this chapter when we shall go on to use it to introduce a further technique of the calculus, the technique of integration. Whereas differentiation is concerned essentially with rates of change, integration is concerned with describing problems involving accumulation. For example, to move an object from one point to another – be it a bag of cement or a bicycle – requires the performance of work. Unlike concepts of rates of change or speed, which are tackled by differentiation and are concerned with a particular instant at time, an idea such as 'total work done' is essentially to do with an accummulation over a period. Another example might be to calculate the total amount of water flowing through a pipe over a given period of time. We have seen that we have a useful geometric interpretation of the derivative – whatever it actually represents in a particular practical case, it can be envisaged in terms of the slope of a tangent to a curve. In the same way, you will see in this chapter that integration can always be thought of in terms of area, whatever the particular practical situation. It is quite remarkable that many different problems can in a sense be brought together like this and thought of in common terms, by interpreting them as an area.

First, though, let's tidy up our ideas on limits.

Two examples we have already met illustrate the idea of a limit very well. In the very first chapter we met the sequence of numbers

$$1, 2, 1.5, 1.666, 1.6, 1.625, 1.6154, \ldots$$

in connection with the Fibonacci sequence, and in Chapter 6 we had a sequence of approximations to a speed (Figs. 6.7–6.9):

$$4.57, 5, 5.5.$$

The essential characteristic of both of these examples is that the numbers in each sequence cluster closer and closer together as one moves further along the sequence. As we remarked in Chapter 5 when we were discussing the real numbers, this means that in each case there is some number, the *limit* of the sequence, about which this clustering takes place. The outstanding issues are to say exactly what we mean by 'clustering' and exactly what we mean by 'limit'.

So, what is a 'limit'?

To formulate the definition, we follow a procedure that we have seen frequently before – we translate our ordinary language description of the idea into mathematical symbols. First, what do we mean by 'clustering'? Simply that the numbers bunch closer and closer together. To describe this mathematically we need a way of expressing in symbols the 'distance' apart of two numbers. This 'distance' can be interpreted just as the difference of the two numbers: so that the 'distance' between 5 and 3, for example is just 2, that is to say $5-3$. And this ties in very nicely with the interpretation of numbers on a number line.

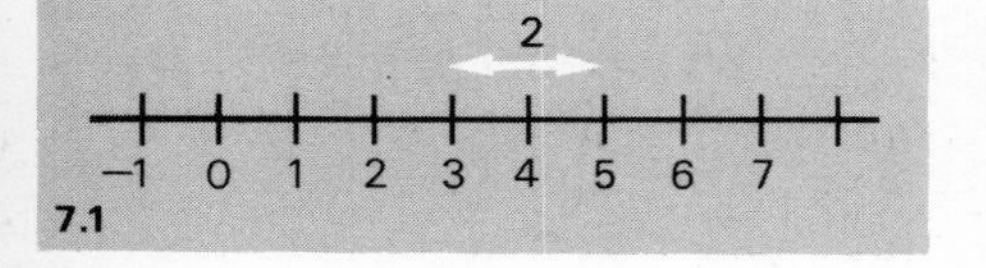

7.1

Measuring the closeness of two numbers

There is one sophistication that we need. If we want to talk of the 'distance' between two numbers, x and y, say, do we work out $x-y$ or $y-x$? This is a genuine difficulty because, even if we decided on a convention – say to take $y-x$ – two problems arise. In the first place, this number may turn out to be negative, and to talk of the distance between two numbers as being negative does not make much sense. Secondly, if the distance between x and y is $y-x$, then presumably the distance between y and x is $x-y$ – but surely they should be the same. We can get round both problems if we take the distance to be $y-x$ if y is bigger than x and $x-y$ if y is smaller than x. We have already seen a way of writing this on p. 44, where we used the modulus sign, $|x-y|$ to denote the numerical magnitude of a number. So, for example $|-4|$ is just 4 and $|x-y|$ is $x-y$ if x is bigger than y, and $y-x$ if y is bigger than x.

So we now have a way of saying how close together two numbers x and y are – the smaller $|x-y|$ is, the closer together they are. Let's apply this idea to sequences.

Convergence of sequences

Suppose we label the first term in a sequence by a_1, the second by a_2, the third by a_3, and so on. We have to use symbols like this, because we want to argue in general terms – to talk about any sequence. But don't be put off. The letters a_1, a_2, a_3, and so on just stand for numbers, so if you start to get confused, just think of them as numbers; it should help. So the sequence is

$$a_1, a_2, a_3, \ldots .$$

Let's denote two arbitrary integers (whole numbers) by m and n – then a_m and a_n are two arbitrary numbers in the sequence; the mth and nth numbers along.

$$a_1, a_2, a_3, \ldots, a_m, \ldots, a_n, \ldots .$$

As we have seen, we can measure the closeness of a_m to a_n by the number $|a_m - a_n|$. If the numbers in the sequence are clustering closer and closer together as we move further along the sequence, then the number $|a_m - a_n|$ is getting smaller as m and n get larger and larger.

This gives us a condition for a sequence of numbers to *converge* – the number $|a_m - a_n|$ must get smaller and smaller as m and n get larger and larger.

Although this condition is certainly necessary to ensure that a sequence converges, it is not sufficient. Consider for example the sequence

$$\tfrac{3}{4}+\tfrac{1}{8}, \tfrac{1}{4}-\tfrac{1}{8}, \tfrac{3}{4}+\tfrac{1}{16}, \tfrac{1}{4}-\tfrac{1}{16}, \tfrac{3}{4}+\tfrac{1}{32}, \tfrac{1}{4}-\tfrac{1}{32}, \ldots .$$

It looks rather complicated, but it is designed especially to make a point. You can get a feel for the sequence by thinking of it geometrically – on the number line. The first two terms, $\frac{3}{4}+\frac{1}{8}$ and $\frac{1}{4}-\frac{1}{8}$, are respectively halfway between B and C and O and A. The next two terms, $\frac{3}{4}+\frac{1}{16}$ and $\frac{1}{4}-\frac{1}{16}$, are halfway between E and C and D and A respectively. And so it goes on. Alternate terms in the sequence approach closer to B or A respectively – halving the distance from B or A at each stage.

7.2

Can you see that the terms of the sequence must therefore get closer and closer together, the further we go along the series. In other words, $|a_n - a_m|$ gets smaller and smaller as m and n get larger and larger. And this, you may remember, is the condition that we had for a sequence to be convergent. But this present sequence does not converge – although the terms get closer together, they do not 'cluster' together, because successive terms are never closer together than $\frac{1}{2}$, the distance between $\frac{3}{4}$ and $\frac{1}{4}$.

The point of this example is to demonstrate the inadequacy of the condition that $|a_n - a_m|$ should get smaller for guaranteeing the convergence of a sequence. It is certainly a necessary condition – every convergent sequence must have this property – but it is not sufficient: not every sequence with this property is convergent. If we seek a property that is both necessary and sufficient, then we need something rather stronger. The weakness in our present condition is that it describes only the property of terms getting closer together, but not the idea of *clustering* together.

We can meet this second requiring that $|a_n - a_m|$ not only gets smaller but can be made *as small as we please* by going far enough along the sequence. This condition is rather like the rules in a competition. If I claim that a sequence

$$a_1, a_2, a_3, \ldots$$

is convergent, then I must show that whatever number, d, I am given – however small – I can find a place in the sequence so that if a_m and a_n are beyond that point then $|a_n - a_m|$ is less than d. The smaller the number d that I am given, the further along in the sequence I would have to go. This is simply because the closer together you want the terms to be, the further along the sequence one has to go.

Do you notice the subtlety in the way this condition has been devised? The difference between this one and our earlier one is simply that not only does $|a_n - a_m|$ become smaller but, under the improved condition, it can be made as small as we please. And this small modification does the trick – the new condition is both necessary and sufficient for the sequence to converge. So far, so good; what we have done is to work out a condition that guarantees the convergence of sequence, so ensuring that a limit exists. What we now have to do is to actually say what that limit is.

Limits

Let's take it a step further. Suppose we have a convergent sequence. If we have, we know that the quantity $|a_n - a_m|$ can be made as small as we like by taking m and n large enough. This is the mathematical expression of the 'clustering' property and it leads us to a new concept – a way of describing the number about which the terms of the sequence are clustering. This number is called the *limit* of the sequence, and we define it by using exactly the same ideas as we have been, but in a slightly different way.

If the number l is the limit of a sequence $a_1, a_2, a_3, \ldots$, then the quantity

$$|a_n - l|$$

can be made as small as we please for every term a_n beyond some point in the sequence. Put rather more graphically, given any number, d, however small, then by going far enough along the sequence a term (we can call it a_N) can be found such that from that point onwards every term in the sequence is within a distance d from the limit, l.

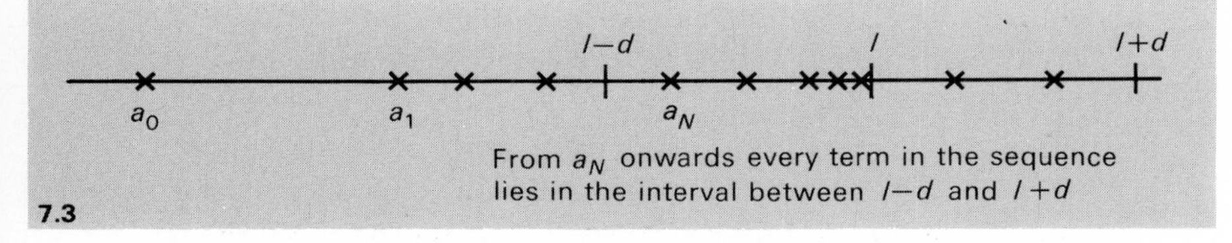

From a_N onwards every term in the sequence lies in the interval between $l-d$ and $l+d$

7.3

The smaller the value of d, the larger is the value of N likely to be: in other words, the closer you want to get to l, the further along the sequence you are likely to have to go.

Back to derivatives

I promised in the previous chapter to show you how this idea of a limit helped with the problems that beset the calculus on its introduction. Let us just remind ourselves what those difficulties were.

The crux of the definition of the derivative of a function, f, is the expression.

$$\frac{f(x+h)-f(x)}{h}$$

and its behaviour as h approaches zero.

The difficulty with this idea is that division by zero is not defined, and so the expression

$$\frac{f(x+h)-f(x)}{h}$$

does not make sense at precisely the point that we want it to: when h is zero both the numerator and the denominator are zero. Newton's attempt to calculate the 'ratio of two vanishing quantities' was fiercely criticised by Bishop Berkeley in a book *The analyst: addressed to an infidel mathematician.* The vehemence of the attack may not have been entirely divorced from the fact that Newton had seduced one of the Bishop's pupils from a study of theology to a pursuit of mathematics, but the critique was nevertheless a brilliant piece of work. Newton was not unaware of the difficulties but the criticisms were not satisfactorily countered until late in the nineteenth century. Fortunately the intervening period was not wasted: without waiting for a more satisfactory foundation to be constructed the calculus was applied with outstanding success to a wide variety of problems, and made a vital contribution to the rapid growth in scientific and technological knowledge that took place during the eighteenth and nineteenth centuries. This period of mathematical history has been admirably described recently by Nicholas Bourbaki.

THE

ANALYST;

OR, A

DISCOURSE

Addreſſed to an

Infidel MATHEMATICIAN.

WHEREIN

It is examined whether the Object, Principles, and Inferences of the modern Analyſis are more diſtinctly conceived, or more evidently deduced, than Religious Myſteries and Points of Faith.

By the AUTHOR of *The Minute Philoſopher.*
Bp. Berkeley

Firſt caſt out the beam out of thine own Eye; and then ſhalt thou ſee clearly to caſt out the mote out of thy brother's eye. S. Matt. c. vii. v. 5.

LONDON:
Printed for J. TONSON in the *Strand.* 1734.

> In 1604, at the height of his scientific career, Galileo argued that for a rectilinear motion in which speed increases proportionally to distance covered, the law of motion should be just that which he had discovered in the investigation of falling bodies ($x = ct^2$). Between 1695 and 1700 not a single one of the monthly issues of Leipzig's *Acta Eruditorum* was published without articles by Leibniz, the Bernoulli brothers, or the Marquis de l'Hôpital treating, with notation only slightly different from that which we use today, the most varied problems of differential calculus, integral calculus and the calculus of variations. Thus, in the space of almost precisely one century infinitesimal calculus or, as we now call it in English, The Calculus, the calculating tool par excellence, had been forged; and nearly three centuries of use have not completely dulled this incomparable instrument.

As we have mentioned the justification of most of these applications lay in the notion of 'limit'. But we have seen limits introduced in the context of sequences, so how do we apply it to the expression

$$\frac{f(x+h)-f(x)}{h}?$$

What we do is this. When we think of h approaching closer and closer to zero, we think of it doing so by taking the values of a sequence that has limit zero. For example, h could take on, in turn, the values

$$1, \tfrac{1}{2}, \tfrac{1}{4}, \ldots.$$

For each of these values of h, you can work out the value of $\dfrac{f(x+h)-f(x)}{h}$ and, if x is assigned a particular numerical value, then the expression will yield another sequence of numbers. Suppose, for example

$$f(x) = \frac{1}{x},$$

then

$$f(x+h) = \frac{1}{x+h}$$

and so

$$\frac{f(x+h)-f(x)}{h} = \frac{\dfrac{1}{x+h} - \dfrac{1}{x}}{h}.$$

Now let's give x a value, say $x = 4$. Then the expression becomes

$$\frac{\dfrac{1}{4+h} - \dfrac{1}{4}}{h}.$$

Next, give h successively the values of the sequence

$$1, \tfrac{1}{2}, \tfrac{1}{4}, \tfrac{1}{8}, \ldots$$

so that we put $h = 1$ and work out the value of $\dfrac{\dfrac{1}{4+h} - \dfrac{1}{4}}{h}$, then try it with $h = \frac{1}{2}$, and so on. This produces the sequence

$$\frac{\frac{1}{4+1} - \frac{1}{4}}{1} = -0.050,$$

$$\frac{\frac{1}{4+\frac{1}{2}} - \frac{1}{4}}{\frac{1}{2}} = -0.056,$$

$$\frac{\frac{1}{4+\frac{1}{4}} - \frac{1}{4}}{\frac{1}{4}} = -0.059,$$

and so on. So we get a sequence of numbers -0.050, -0.056, $-0.059, \ldots$. If this sequence converges, it has a limit. We call this limit the limit as h approaches 0 of

$$\frac{f(4+h)-f(4)}{h}.$$

And this is the value of the derivative of f, worked out at the point 4, just as $\dfrac{f(x+h)-f(x)}{h}$ is the value of the derivative of f at the point x. So the limit is $f'(4)$. We write

$$f'(4) = \lim_{h \to 0} \frac{f(4+h)-f(4)}{h},$$

where we read the symbols '$\lim\limits_{h \to 0}$' as 'the limit as h approaches zero'. Before we go on, let's just tidy up one point. To arrive at our extension of the idea of a limit, we chose a particular sequence of values of h; the sequence

$$1, \tfrac{1}{2}, \tfrac{1}{4}, \tfrac{1}{8}, \ldots.$$

We chose this because it is a particularly straightforward sequence. We could have chosen any sequence. This is no idle comment, because it is important that we get a 'proper' answer for the limit. In fact we say that $f'(x)$ is 'properly defined' only if whatever sequence of values we take for h having limit zero then the sequence of numbers $\dfrac{f(4+h)-f(4)}{h}$ converges to the same limit – it is that limit we call $f'(4)$.

It is not a point that you need worry about – we mention it only for the sake if accuracy (or to ease a methematician's conscience!). It is interesting to note, however, that the idea of a limit was precisely what we needed to specify the complete real-number system. Now we see that the very same idea removes an important difficulty in the calculus.

Calculating areas

As we have mentioned earlier, the idea of differentiation (which is concerned with rates of change) is one of the two key ideas of the calculus. We have now seen how that notion is dependent on the concept of limit. What we are going to do now is to apply this notion of limit to develop the other key idea – that of *integration.*

Just as the technique of differentiation can be used to tackle problems of many different types, so integration has an amazing ubiquity. But just as the speciality of the house for differentiation is problems involving motion, so, in much the same sense integration has as its forte problems involving the calculation of areas.

The similarity between the two situations is particularly interesting because just as differentiation actually helps us to *understand* what we mean by velocity, so does integration help us understand what we mean by area. You might well say that you don't need any help in understanding what is meant by area, but are you sure? What do you make of the areas contained within the following figures?

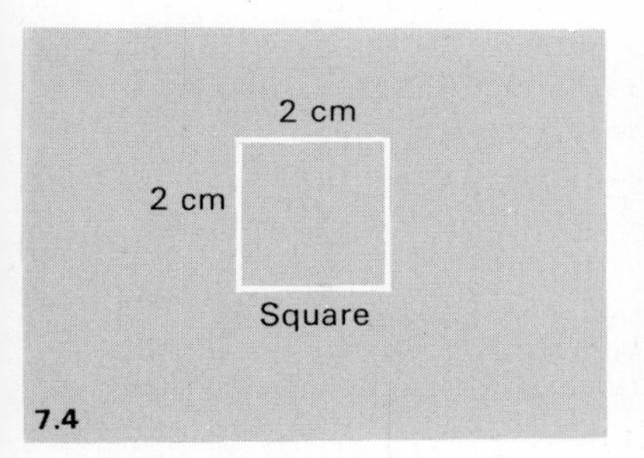

7.4

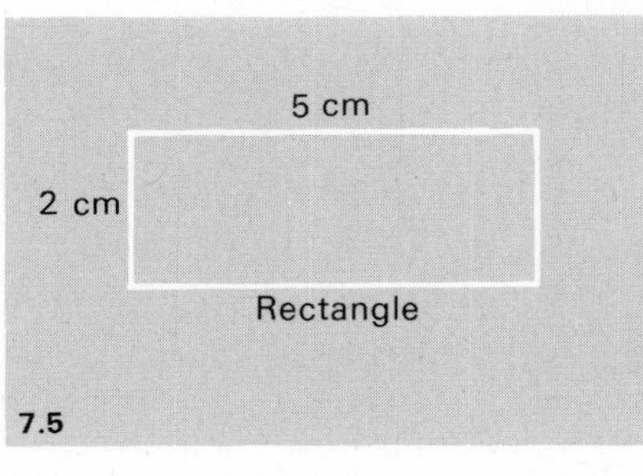

7.5

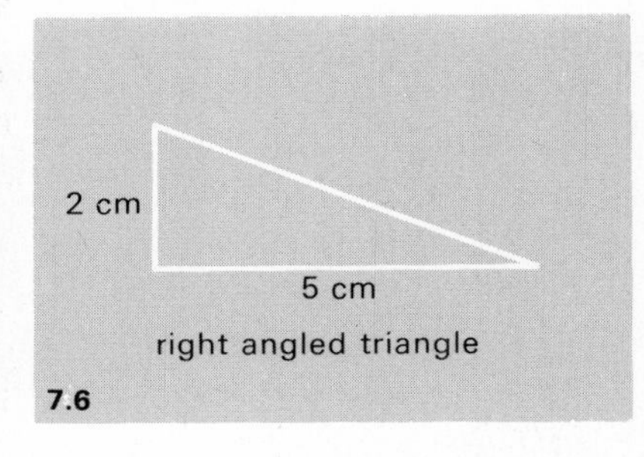

7.6

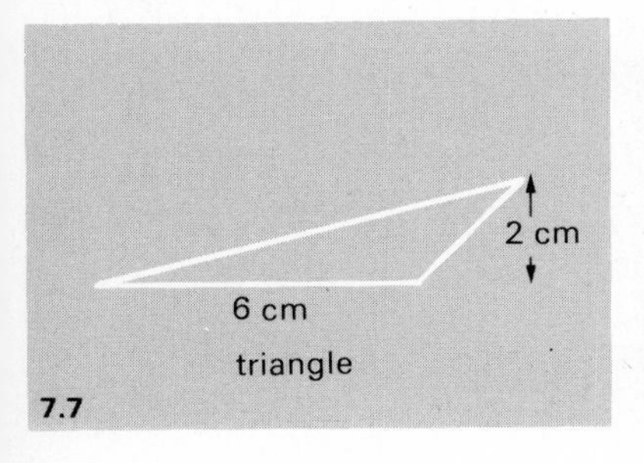

7.7

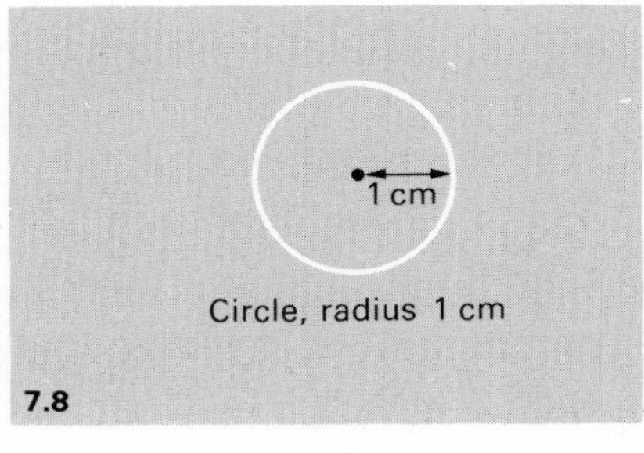

7.8

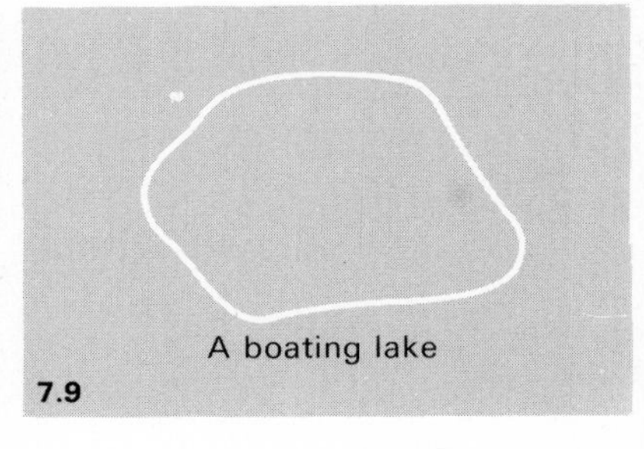

7.9

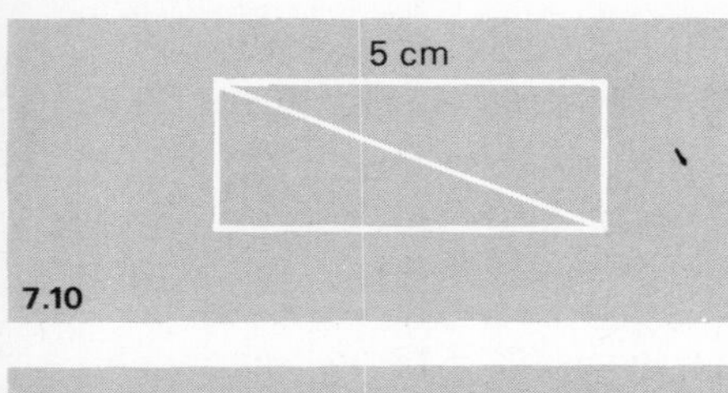

7.10

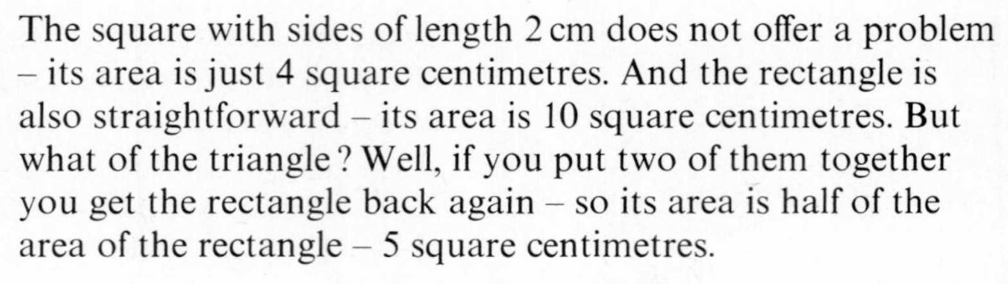
The square with sides of length 2 cm does not offer a problem – its area is just 4 square centimetres. And the rectangle is also straightforward – its area is 10 square centimetres. But what of the triangle? Well, if you put two of them together you get the rectangle back again – so its area is half of the area of the rectangle – 5 square centimetres.

7.11

This may seem fairly straightforward to you, but it is in fact quite a sophisticated step. Think what is actually meant by saying that the rectangle has area 10 square centimetres. What we mean is that 10 squares of side 1 cm can fit exactly into the rectangle.

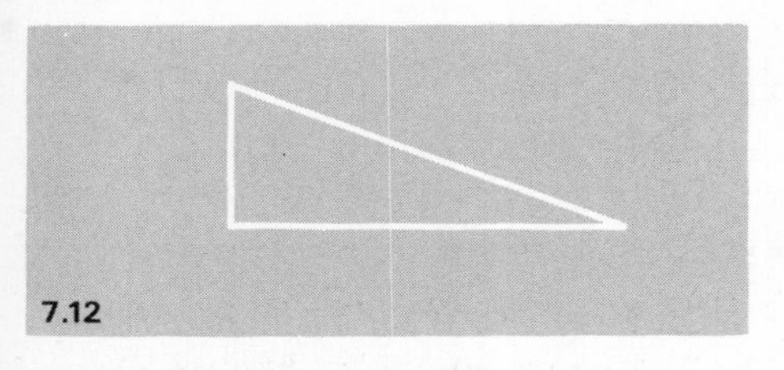
7.12

But there is no way that we could fit 5 such squares into the triangle.

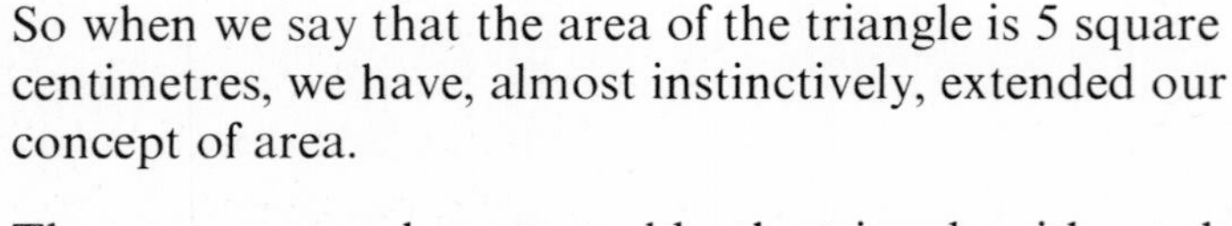
So when we say that the area of the triangle is 5 square centimetres, we have, almost instinctively, extended our concept of area.

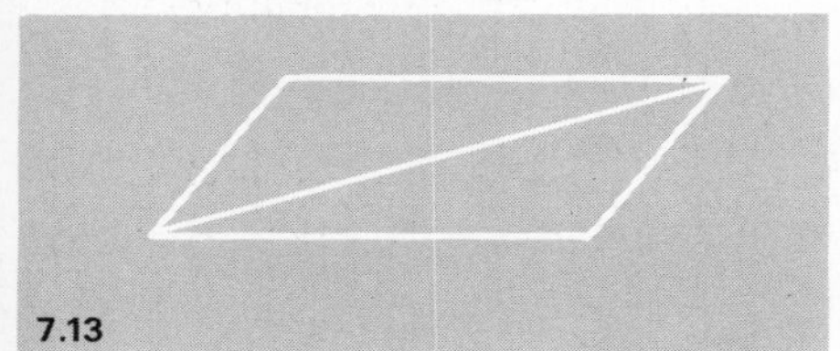
7.13

The next area we have to tackle, the triangle without the right-angle, presents us with no new problem of principle: it just requires a little cunning.

This triangle has half the area of a parallelogram, but how do we find the area of the parallelogram? Like this:

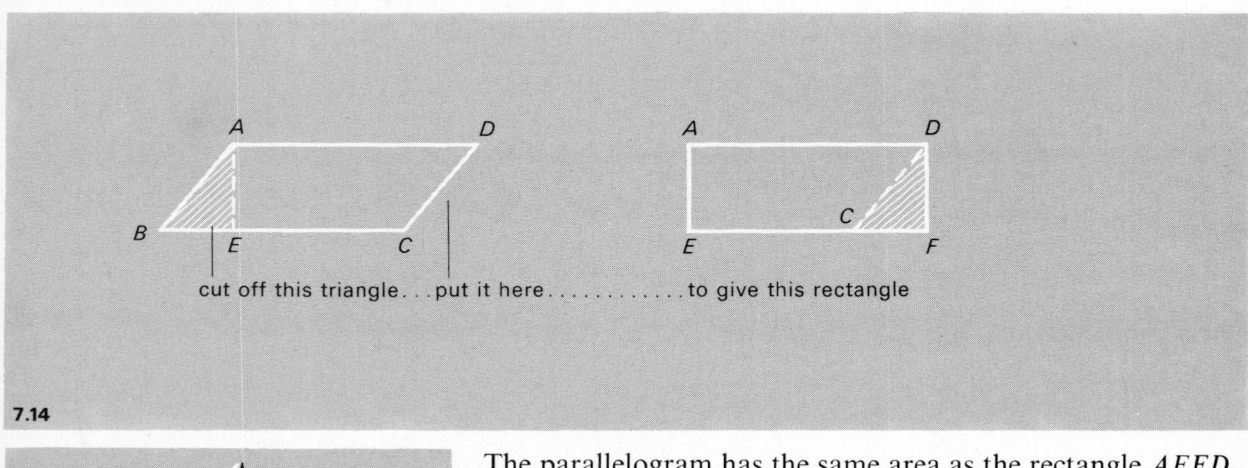

7.14

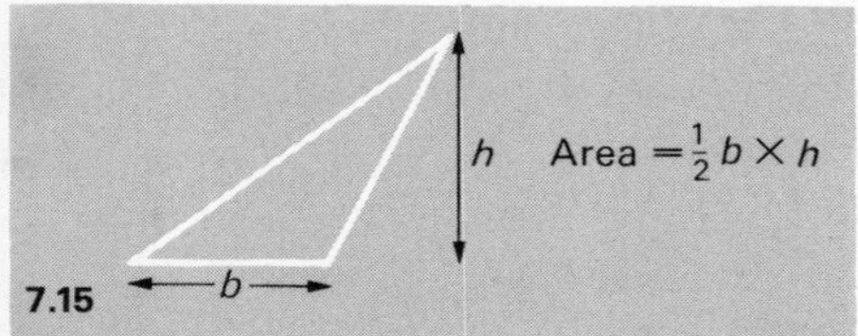

7.15

The parallelogram has the same area as the rectangle $AEFD$, which is just 24 square centimetres. So the triangle has area 12 square centimetres.

The same process could be applied to any triangle to give a formula for its area.

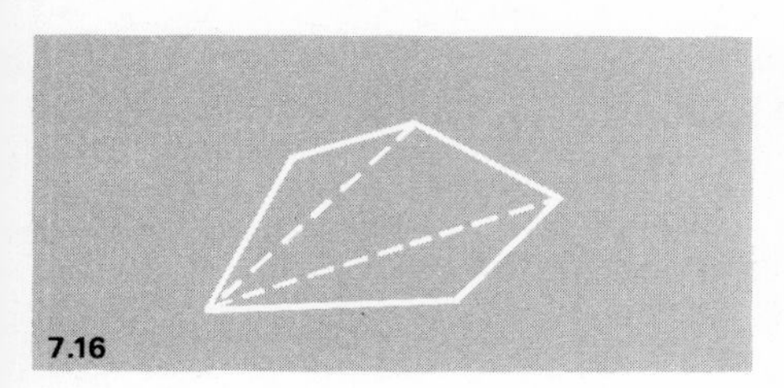

7.16

You can probably see that any region bounded by straight lines can be coped with because it can be split up into triangles.

But what about a region with a curved boundary? Obviously that cannot be split into triangles. Take, for example a circular region. One way to tackle it is like this.

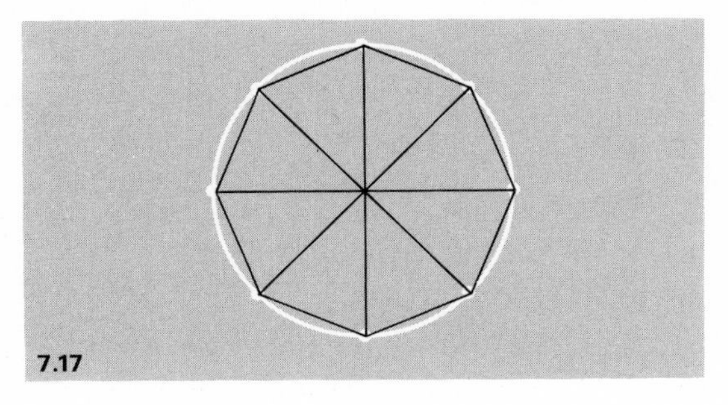

7.17

The sum of the triangular areas is an approximation to the circular area – and that is the key to the approach. If we use more and more triangles, our approximation gets better and better.

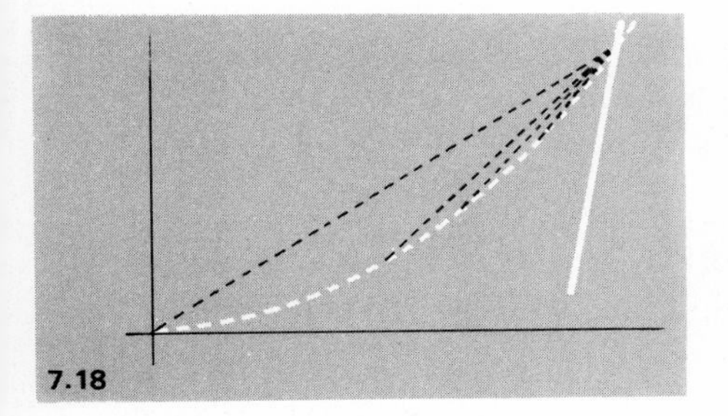

7.18

Does the idea ring any bells? Do you recognize a similarity with the way we got at the notion of a tangent by successive approximations on p. 158?

In fact, although the technical manipulations that are required are a little beyond us at the moment, this is precisely the way in which a formula for the area of a circle is obtained – and it requires exactly the same notion, the idea of a limit.

Let's be a little more precise. If the angle at the centre of the circle is divided into n equal parts, we get n equal triangles. Let's denote the total area of these triangles by S_n. Then the larger the value of n we take, the smaller is the angle and more triangles we get, and the better is the approximation.

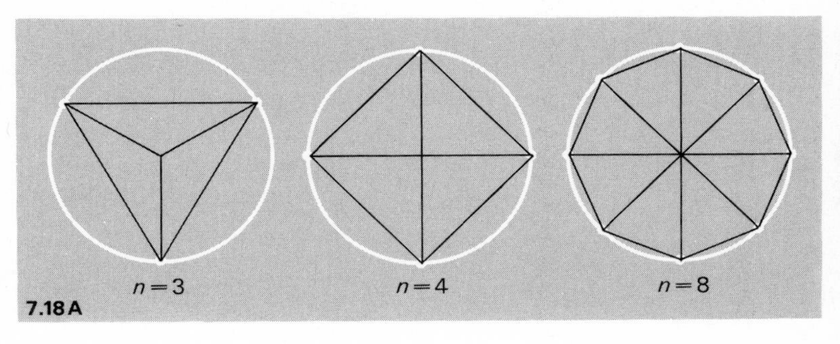

7.18 A

When the value of S_n is worked out, the result is an expression in n and we take the limit of the sequence S_n as the area of the circle.

The idea of a limit is used not just to calculate an area, but to *define* it.

There is far more to this last step than may meet the eye. The notion of limit not only serves to express how the approximations get closer and closer to the area, but it actually serves to *define* what we mean by the area of a region with a curved boundary. You see, we measure area in units, square centimetres, or whatever, and this is indicative of the fact that to begin with we define only the area of square – which is then generalized, as we have seen, to rectangles, triangles, and then any figure with straight sides. To extend the idea to regions with curved boundaries requires the much more sophisticated idea of a limit.

So, as far as areas are concerned, limits have two roles: they are used to define areas with curved boundaries and they provide a technique for calculating the areas so defined.

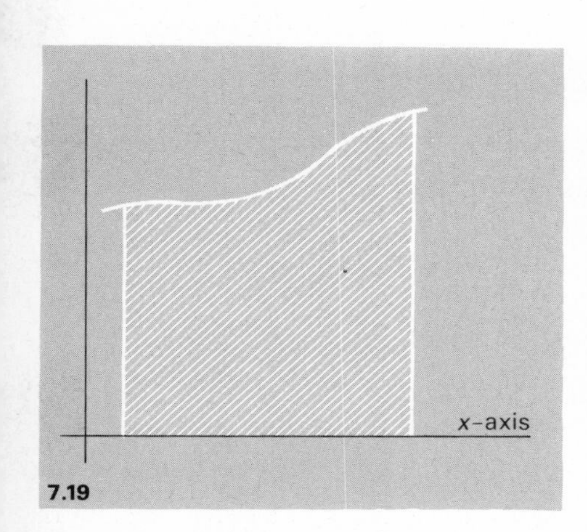

7.19

Areas under graphs

Apart from familiar shapes like circles, the most frequently arising shapes with curved boundaries are those whose boundary, or at least part of it, is formed by the graph of a function, like the shaded region in Fig. 7.19.

We approach this problem in a similar way to the area of a circle and, in doing so, we shall arrive at the second major ingredient of the calculus – the integral.

What we do this time is to split the area up into rectangles.

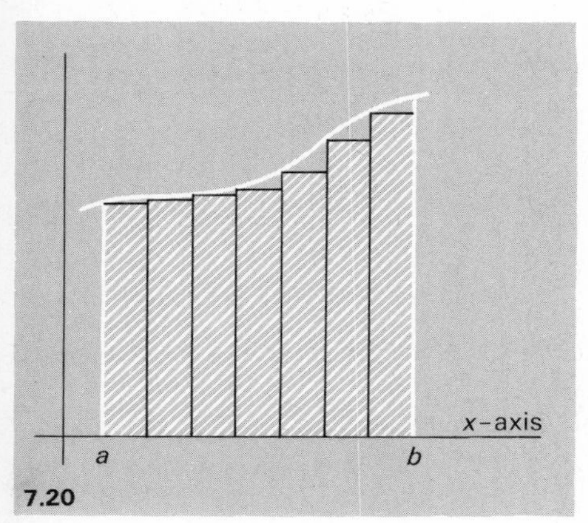

7.20

If you look at Fig. 7.20, you will see that what we have done is to approximate the area bounded by the graph of a function f, the x-axis, and the lines drawn at $x = a$ and $x = b$. And we have done it by drawing rectangles, each of width h, with the left-hand side of each rectangle just reaching up to the graph. So the height of the first rectangle is $f(a)$, the height of the second is $f(a+h)$, the third $f(a+2h)$, and so on.

This means that the area of the first rectangle is $h \times f(a)$, the area of the second is $h \times f(a+h)$, and so on. The total area of the rectangles is

$$h \times f(a) + h \times f(a+h) + h \times f(a+2h) + \dots + h \times f(b-h).$$

(By the way, you may have noticed that there is a restriction on the value that we can assign to h: to get an exact division of the interval from a to b, a whole number of rectangles, h must divide exactly into $b-a$, the length of the interval. Indeed, if there are n rectangles, then $h = \dfrac{b-a}{n}$. It is still possible to choose h with considerable freedom, because n can be any positive whole number.)

You can probably see the idea that we are going to develop: the more rectangles we take, the better will the total area of the rectangles approximate to the area that we are trying to find. If we use the symbol S_n to denote the sum of the areas of the rectangles when we have n of them, then we can develop a sequence of approximations

$$S_1, S_2, S_3, \ldots,$$

corresponding to successively thinner rectangles.

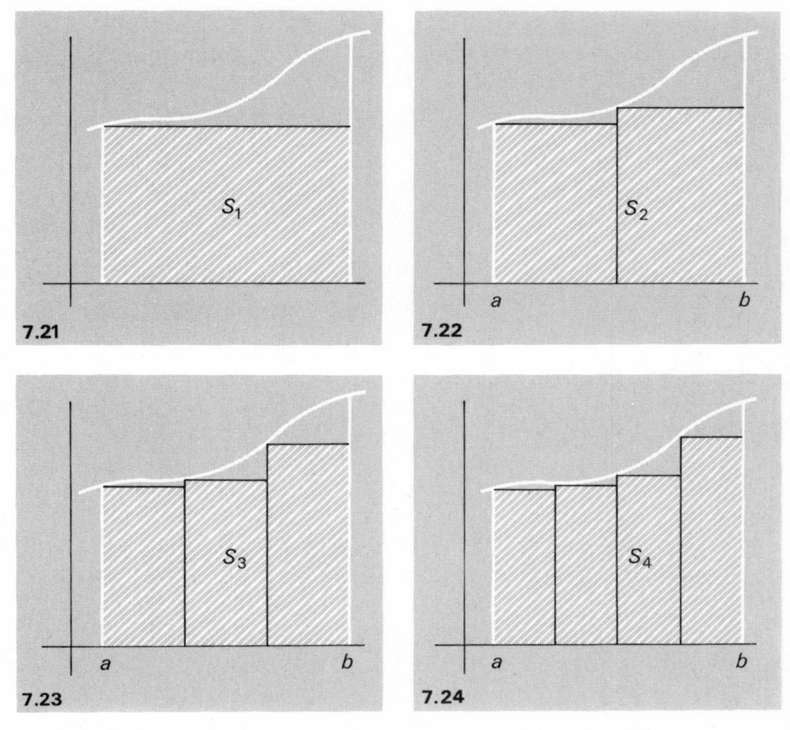

7.21 7.22 7.23 7.24

Now comes the step which may be becoming familiar. Suppose the sequence S_1, S_2, S_3 ... has a limit. Then we take this number as the *definition* of the area in question. The approach is reminiscent of the way we defined the tangent to a curve as the limit of a sequence.

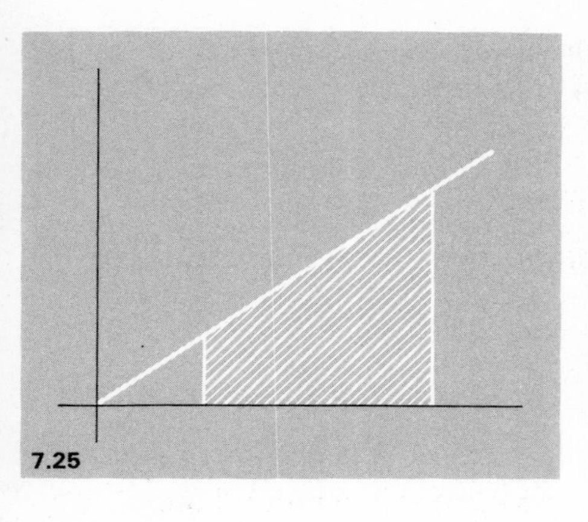
7.25

Let's see how this works out for a simple area, by taking the graph as a straight line. We can take the graph with equation $y = x$.

We already know how to calculate this area – but in a sense that is why it is important. What we are trying to do is to extend our concept of area and our ability to calculate it. It is clearly desirable that the new, more powerful, technique that we develop in this way still gives us the answers we expect for the more elementary cases.

In this particular example, we can calculate the area in question by subtracting the area of the small triangle (unshaded) from the area of the large triangle. The small triangle has area $a^2/2$, the large one has area $b^2/2$, and so the area that we seek is

$$\frac{b^2}{2} - \frac{a^2}{2}.$$

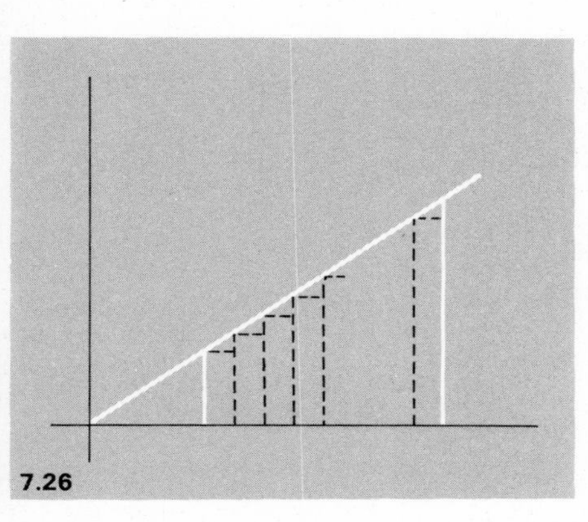
7.26

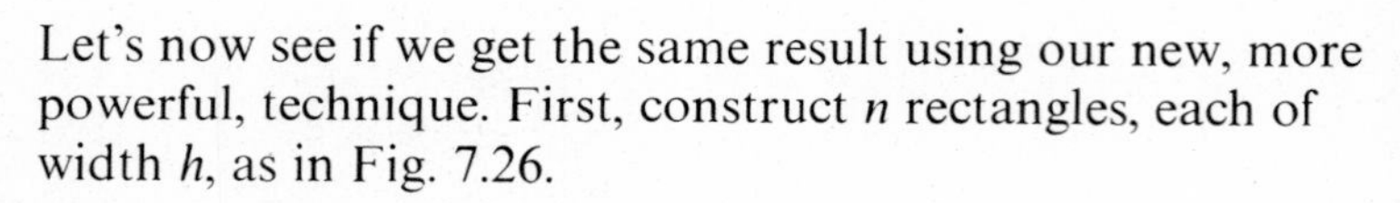

Let's now see if we get the same result using our new, more powerful, technique. First, construct n rectangles, each of width h, as in Fig. 7.26.

Since the graph has equation $y = x$, the height of the first rectangle is a, the height of the second is $a + 1h$, the third $a + 2h$, and so on. The height of the last rectangle is $a + (n-1)h$. So the areas are

$$ha,\ h(a+h),\ h(a+2h), \text{ and so on.}$$

The total area of the n rectangles is

$$ha + h(a+h) + h(a+2h) + \ldots + h(a+(n-1)h).$$

We can rewrite this as

$$h[a + a + h + a + 2h + \ldots + a + (n-1)h].$$

There are n as inside the bracket, so it amounts to

$$h[na + h + 2h + 3h + \ldots + (n-1)h]$$

which, in turn, can be written as

$$h[na + h(1 + 2 + 3 + \ldots + (n-1))]$$

Now look at what is inside the round brackets:

$$1 + 2 + 3 + \ldots + (n-1).$$

We learnt how to cope with this expression way back in Chapter 1 – it works out to

$$\frac{n(n-1)}{2}.$$

So we can now rewrite the total area of the rectangles as

$$h\left[na + \frac{(hn(n-1))}{2}\right].$$

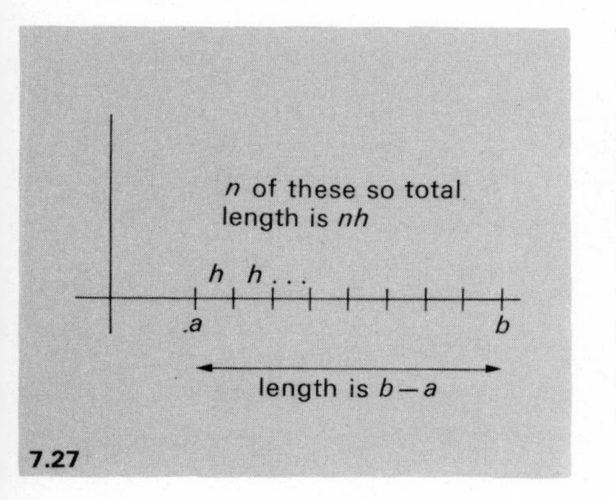

7.27

Now, try to remember what we have to do as the next step: we must let h get smaller and smaller and n get larger and larger, so that we get more and more thinner and thinner rectangles. But we cannot let h get smaller and n get larger independently, because the total width of the rectangles has got to fill out the interval from a to b exactly. In other words we must always have $nh = b-a$. So before we go any further, let's build this into the expression. We can arrange the second term inside the square brackets so that the complete expression can be written as

$$h\left[na + \frac{n^2h-nh}{2}\right],$$

which is simply

$$h\left[na + \frac{n^2h}{2} - \frac{nh}{2}\right].$$

Now let's get rid of the brackets by multiplying each term inside the brackets by the h outside. This gives

$$hna + \frac{n^2h^2}{2} - \frac{nh^2}{2}.$$

Now we are in a position to use the fact that $nh = b-a$; it turns our expression into

$$a(b-a) + \frac{(b-a)^2}{2} - \frac{(b-a)h}{2}.$$

What happens to this as h gets smaller and smaller? Nothing happens to the first two terms, but the third term, $\frac{(b-a)h}{2}$, must get smaller and smaller.

So what is the limit of this expression as h approaches zero? It must be just

$$a(b-a)+\frac{(b-a)^2}{2},$$

and that is the answer we get for the area.

But if you look back to the answer we got for the area by direct calculation, you will see we had

$$\frac{b^2}{2}-\frac{a^2}{2}.$$

It would appear that we may have a little trouble – our answers should surely be the same. But let's not panic; look again at our new answer

$$a(b-a)+\frac{(b-a)^2}{2}.$$

This can be written as

$$ab-a^2+\frac{(b^2-2ab+a^2)}{2}$$

which amounts to

$$ab-a^2+\frac{b^2}{2}-ab+\frac{a^2}{2}.$$

The two terms ab and $-ab$ cancel each other and the terms $-a^2$ and $+\frac{1}{2}a^2$ total to $-\frac{1}{2}a^2$. So finally we get

$$\frac{b^2}{2}-\frac{a^2}{2},$$

the result we sought.

Well, that was a lengthy and rather arduous process. Nobody in his right mind would suggest that this would be the best way to calculate the particular area in question, but remember that was not our purpose. The technique that we have been using is devised for calculating areas for which no other method is available – we were simply illustrating the use in a familiar context, and confirming that the more general technique produces the expected results in the simple cases.

Nevertheless, the technique is a difficult one to apply: the calculation of the limit is possible to perform for only a few functions. But this difficulty is overcome by using another result, whose importance can hardly be overestimated. It is the cornerstone of the calculus and this is reflected in its name, the Fundamental Theorem of Calculus. As the last major item in this book, I want to show you what this theorem is and give some justification for its validity. This is, as we have just said, a major result in mathematics, and if you can go some way to understanding it then there can be no doubt that you have achieved something worthwhile.

A new function

What the Fundamental Theorem does is to link together the two basic ideas of the calculus – calculating rates of change (the province of differentiation) and calculating areas (integration). Before we can explain what the link is, we have to tidy up our work on areas just a little.

You may remember that once we had developed the technique of calculating rates of change, we extended the notion to define a new function. Given a function f, we defined a new function, f', called the *derived function*, where $f'(x)$ is the gradient of the graph of f, evaluated at x. Have a quick look at p. 166 to refresh your memory.

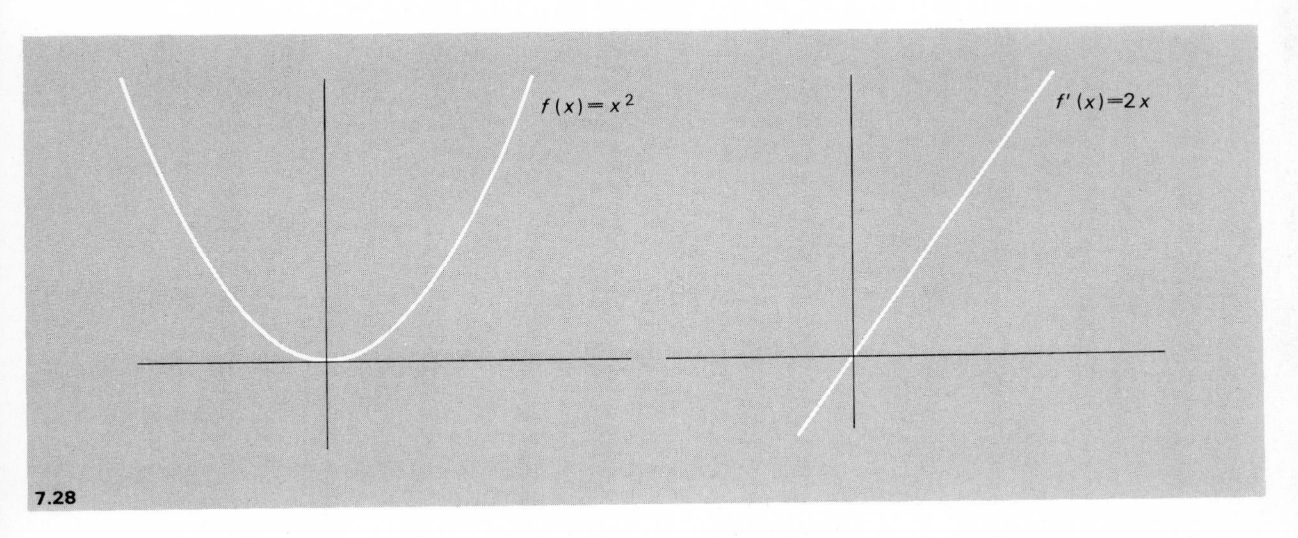

7.28

We can define another new function using our new concept of area. We shall do it in two stages. First, just a question of terminology. The process of evaluating the limit of the sum of the rectangle is called *integration*. The area under the graph of a function, f, between the values

7.29

a and b – the shaded area in Fig. 7.29 is called the integral of the function f between a and b and is denoted by $\int_a^b f$. This may seem a rather strange symbol, but the $\int$ sign can be thought of as an elongated S, standing for a special sort of summation which recalls the summing of the areas of the rectangles in the definition of integration.

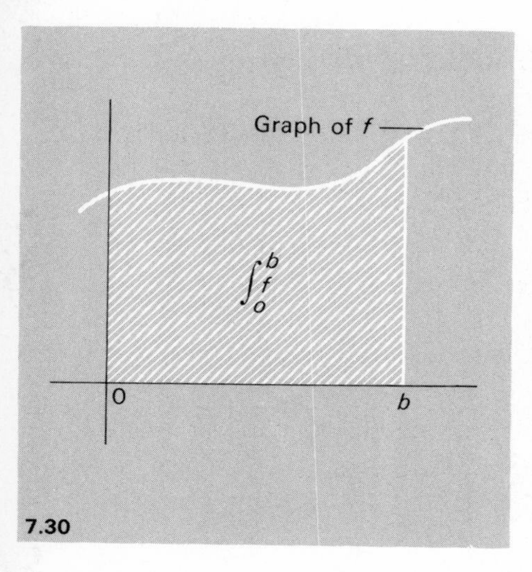

7.30

Now, the value of $\int_a^b f$ must depend not just on the particular choice of the function f, but on the values of a and b. Suppose we fix one of them and make it equal to zero. This gives us $\int_0^b f$.

If you have got that idea you're ready for the next move – which is subtle and sophisticated, but no more so than a similar tactic that we have already employed when discussing differientation.

As b varies, than so does $\int_0^b f$. What this means is that to any particular value of b we can assign another value, simply $\int_0^b f$. We have within our grasp a new function, F, where

$$F(b) = \int_0^b f.$$

For example, we have already seen (on p. 190) that when f is the function where

$$f(x) = x$$

then

$$\int_a^b f = \frac{b^2}{2} - \frac{a^2}{2}.$$

So in this case the function F is specified by

$$F(b) = \frac{b^2}{2}.$$

It is quite convenient to use the symbol x instead of b here, and to write for the function F,

$$F(x) = \frac{x^2}{2}.$$

We shall see why very shortly.

So now we have two ways of constructing new functions. Given a function f, we have two new functions: the *derivative*

$f'(x) =$ gradient of graph of f at x which measures the rate of change of $f(x)$ as x increases

and the *integral*

$$\int_0^x f = \text{area under graph of } f \text{ between } 0 \text{ and } x.$$

We have already listed the derivatives of a few functions (on p. 167) and we have just seen that when $f(x) = x$, $F(x) = \frac{x^2}{2}$.

This is sufficient to illustrate the result we wish to discuss. Look at the pair of functions

$$f(x) = x, \; F(x) = \frac{x^2}{2}.$$

If you look at the list of derivatives on p. 167, you will see that if we differentiate the function F we get

$$F'(x) = x,$$

in other words,

$$F'(x) = f(x).$$

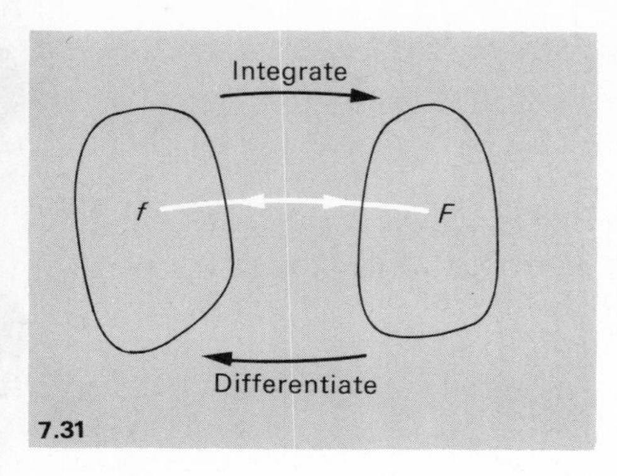

7.31

Do you see what has happened? We start with the function f, integrate it to get a new function, differentiate it again, and what do we end up with? Our original function, f.

In terms of our particular example, we started with the function f where $f(x) = x$ and saw that the area under its graph, between 0 and x is $\frac{x^2}{2}$ and so specifies the function F, where $F(x) = \frac{x^2}{2}$.

And if we differentiate F, that is to say work out the rate of change of F, we get back to f again.

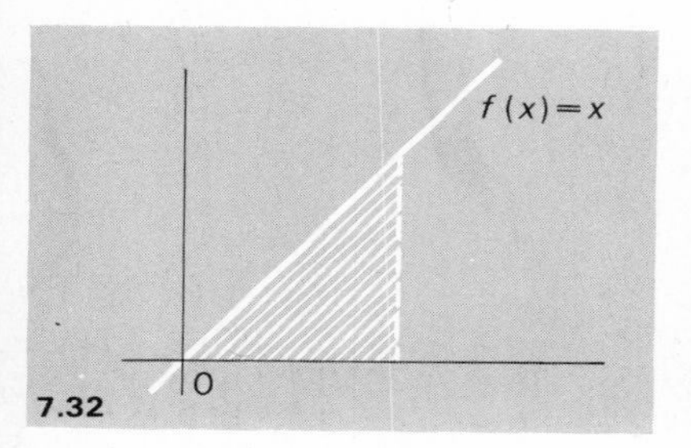

7.32

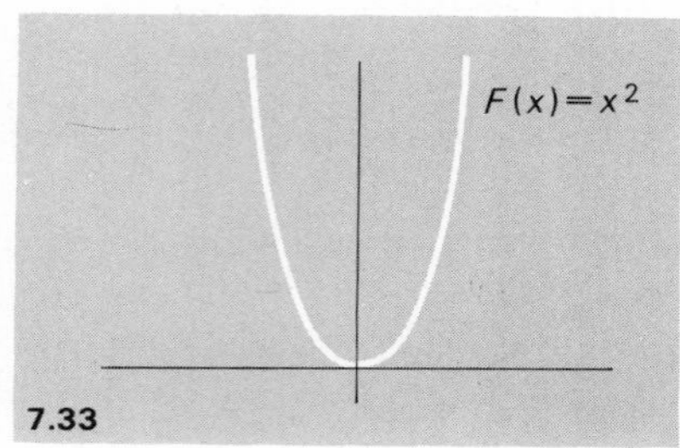

7.33

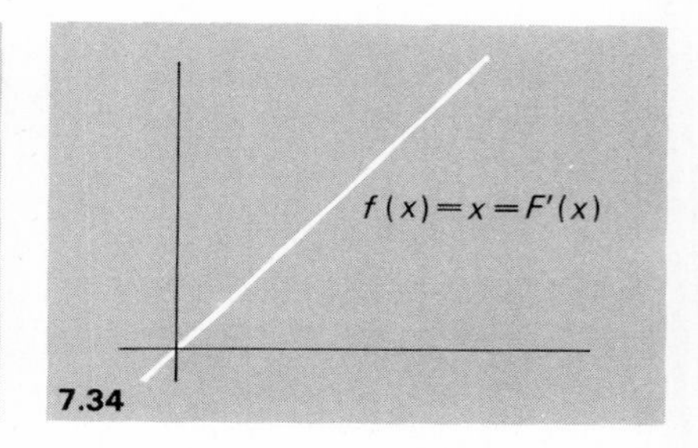

7.34

The Fundamental Theorem of Calculus

Now, just one example of a phenomenon is no real indication of anything remarkable but this particular property is not special to the function that we chose. It always works. Whenever you integrate a function and then differentiate the result, you get back to the original function. In a sense the processes of integration and differentiation are inverse operations – one undoes the effect of the other.

Shortly, we shall see an important practical application of this relationship between integration and differentiation but first, let's dig a little deeper to see if we can find some justification for the result. Remember how we defined the derivative of a function, back on p. 166. The derivative of a function, F, is defined by

$$F'(x) = \lim_{h \to 0} \frac{F(x+h) - F(x)}{h}.$$

So let's interpret the expression

$$\frac{F(x+h) - F(x)}{h}$$

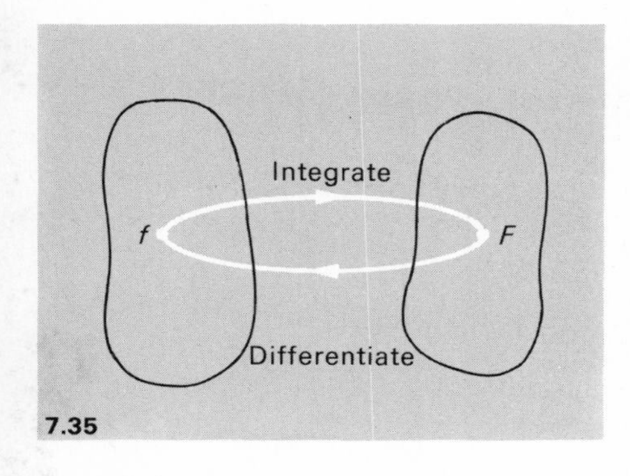

7.35

in the particular circumstances when

$$F(x) = \int_0^x f.$$

Bearing in mind that $F(x)$ is the area under the graph of f between 0 and x, $F(x+h)$ must be the area under the graph between 0 and $x+h$. So $F(x+h)-F(x)$ must be just the area between x and $x+h$.

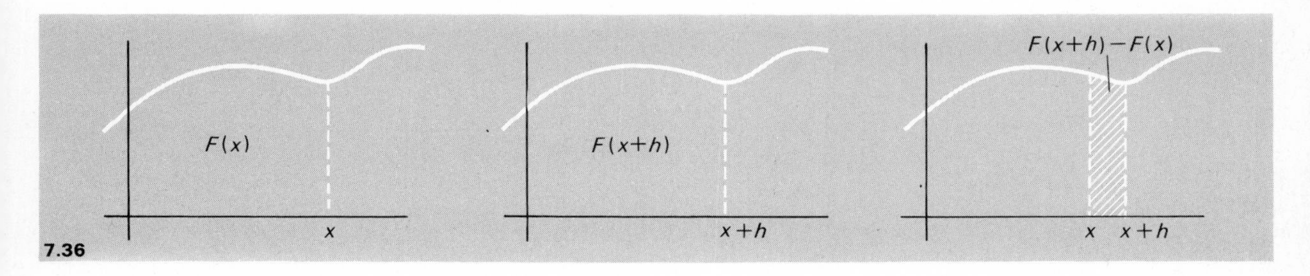

7.36

So how do we interpret $\dfrac{F(x+h)-F(x)}{h}$? Recalling the way we introduced the idea of areas with curved boundaries, we can approximate to the area of the shaded strip by the area of a rectangle. The width of the strip is just h and the height of the left-hand side is just $f(x)$. So an approximation to the area is $hf(x)$.

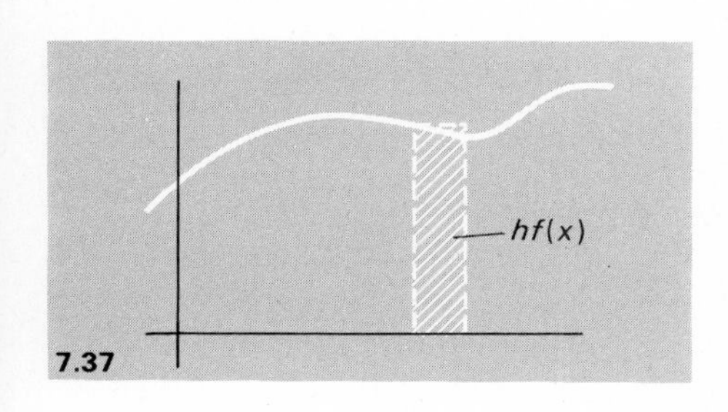

7.37

In other words $F(x+h)-F(x)$ is approximately $hf(x)$. So what is

$$\frac{F(x+h)-F(x)}{h}? \text{ Approximately } f(x).$$

But what do we mean by 'approximately' here – in other words what would improve the approximation? Making the strip thinner, of course. The thinner we make the strip, the better is the approximation and

$$\lim_{h\to 0} \frac{F(x+h)-F(x)}{h} \text{ is just } f(x).$$

We have not actually 'proved' this, but perhaps you will agree that the argument is convincing. In fact a rigorous proof can be supplied and the result that

$$F'(x) = f(x)$$

can be justified. This is the Fundamental Theorem of Calculus.

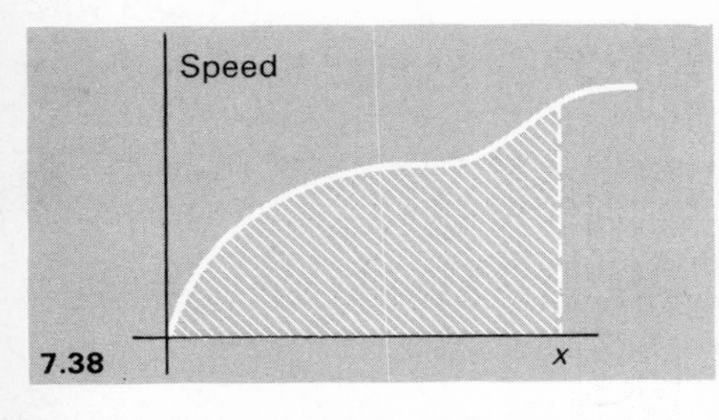

7.38

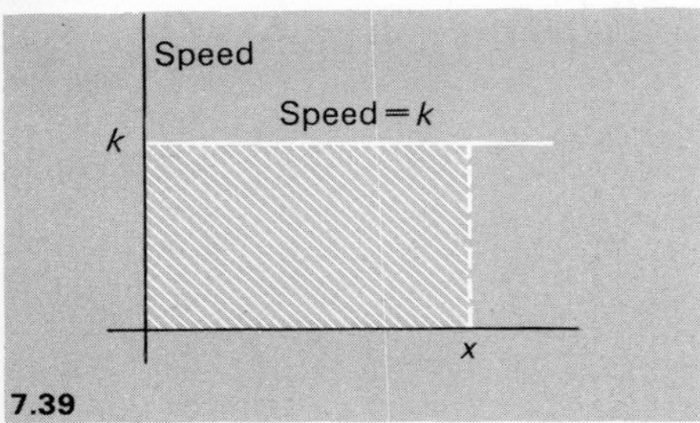

7.39

A practical example

Perhaps a physical example of this result will make it seem less abstract. Suppose $f(x)$ measured the speed of a particle at time x then what is $F(x)$, that is to say what is $\int_0^x f$?

To help you guess, take a simple example for f (this is precisely the way a professional mathematician would work). Suppose $f(x)$ were constant – so we had the situation of a particle moving with constant speed. Let's say $f(x) = k$.

Then $F(x)$ is the area of a rectangle which is kx. And what is the physical interpretation of kx? Remember that this is the product of time of travel (x) with speed (k). If the particle travels at speed k for time x, what is kx? Just the distance travelled!

This is a particular case of a general result that if $f(x)$ represents speed then $F(x) = \int_0^x f$, represents distance travelled.

Now let's interpret the Fundamental Theorem in this context. Remember that the theorem states that

$$F'(x) = f(x).$$

In our particular context, $f(x)$ is velocity and $F(x)$ is distance travelled, and it is indeed the case that $F'(x)$, which is rate of change of distance, is simply the speed.

Evaluating integrals

The next step we want to take is to extract just a little more juice from our result, so that it can be used to develop an important new mathematical technique. Let's just remind ourselves where we have got to: we have seen that if

$$F(x) = \int_0^x f$$

then

$$F'(x) = f(x).$$

We now do something deceptively simple, but rather subtle and cunning. The relation $F'(x) = f(x)$ is one that we have viewed as telling us something about how to differentiate F –

in other words expressing F' in terms of something we know, f. But you can look at the relation the other way round: suppose that by looking at f you were able to spot what F is. Then you would have a technique for working out integrals.

This is because we can replace f by F' and the Fundamental Theorem takes on a new look:

$$\int_0^x F' = F(x).$$

We have written the integral on the left-hand side of the relation to convey more strongly the feeling that we want to get across of moving from the integral to $F(x)$ – in other words a sense of using the result to actually evaluate the integral.

There is just one minor technical difficulty to clear up, but we shall look at that in a moment. First, let's explain in more concrete terms what it is we are trying to get at. Back on p. 167, we had the following table of results about differentiation.

$f(x)$	$f'(x)$
k, any fixed number	0
x	1
x^2	$2x$
x^3	$3x^2$
x^4	$4x^3$
$\frac{1}{x}$	$-\frac{1}{x^2}$
$\frac{1}{x^2}$	$-\frac{2}{x^3}$

For the purpose of illustration take the third result in this list:

$$\text{if } f(x) = x^3, \text{ then } f'(x) = 3x^2.$$

This tells us that if we differentiate the function $x \rightarrow x^3$, we get the function $x \rightarrow 3x^2$. But we now know that differentiation and integration are inverse processes, so we know that if we integrate $x \rightarrow 3x^2$ we get $x \rightarrow x^3$. In other words, the Fundamental Theorem of Calculus enables us to use a table of results in differentiation to get a table of results about integration. Because

knowing that	the theorem tells
if $f(x) = x, f'(x) = 1,$	if $F(x) = 1, \int_0^x F = x$
if $f(x) = x^2, f'(x) = 2x,$	if $F(x) = 2x, \int_0^x F = x^2$
if $f(x) = x^3, f'(x) = 3x^2,$	if $F(x) = 3x^2, \int_0^x F = x^3$

and so on.

The practical importance of this result is that differentiation is a much simpler process, technically, than integration, and so the theorem gets round the difficulty of having to work out extremely difficult limits, and provides a technique for working out the integrals of many, many different functions. Any specialist book on the calculus will show you how to apply the idea in detail.

Tidying up

We mentioned earlier a slight technical difficulty about the result

$$\int_0^x F = F(x).$$

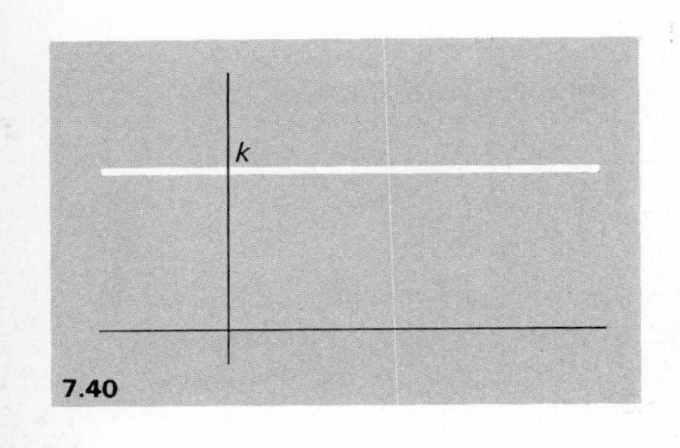

7.40

It is a point that needs clearing up, if only for the sake of accuracy. You may recall that the derivative of a 'constant' function, that is to say one for which $F(x) = k$, where k does not vary with x, is zero. This is obvious, anyway, because the rate of change of a constant is zero! This observation is important in the present context for the following reason. Suppose, for example, we had two functions, f and g, given by

$$f(x) = x^2$$

$$g(x) = x^2 + 3$$

then $f'(x) = 2x$

and $g'(x) = 2x$,

because the constant component, 3, in $g(x)$ differentiates to zero.

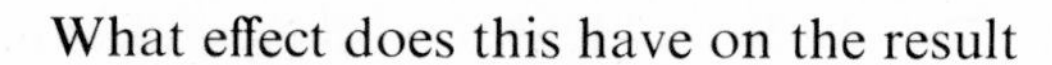

What effect does this have on the result

$$\int_0^x F' = F(x)?$$

Well, it must mean that there is a certain arbitrariness about the choice of F, because whatever function F you might find, that differentiates to the required F', you could add a constant to it and still get F'. So, for example, if one were trying to integrate the function f where $f(x) = 2x$, then it would appear that one could choose any function F of the form $F(x) = x^2 + k$, where k is constant, because any one of them will differentiate to the required function f. What's gone wrong; where does this arbitrariness stem from?

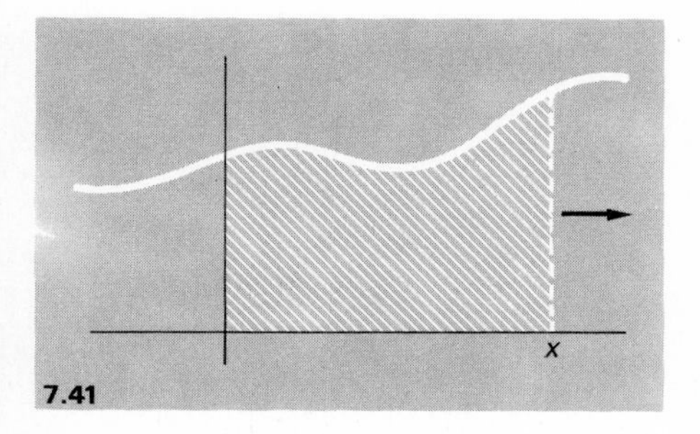

7.41

To answer this, we can go back to the starting point of our discussion of the Fundamental Theorem. We saw earlier that what the theorem was really talking about was the rate of change of the area, $\int_a^x f$, as x increases.

This rate of change of area will depend only on the function, f, and not at all on where you start to measure the area. The *rate of change* of this area

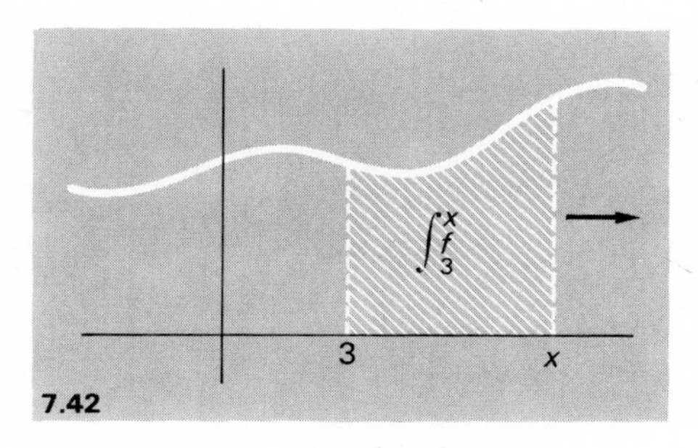

7.42

is precisely the same as this one

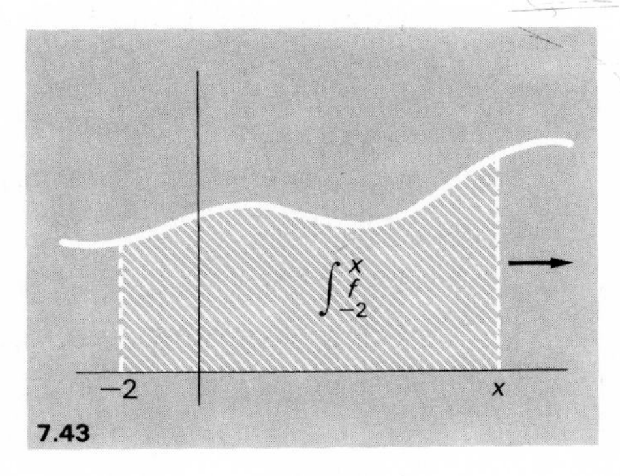

7.43

and, in turn, they are each the same as the rate of change of $\int_a^x f$ whatever value we assign to a. That part of the story is fine; there is no problem about the first part of the fundamental theorem, that if

$$F(x) = \int_a^x f, \text{ then } F'(x) = f(x).$$

The ambiguity arises in taking the result the other way round: we cannot say, without qualification that

$$\text{if } f(x) = F'(x) \text{ then } \int_a^x f = F(x),$$

because $F(x)$ is not precisely defined just by saying that $F'(x) = f(x)$. But we can now see that the ambiguity is not too serious – it arises simply because of the freedom we have to choose the value of a in the expression $\int_a^x f$.

We can, in fact, express this freedom precisely by considering the difference between say

$$\int_a^x f \text{ and } \int_b^x f,$$

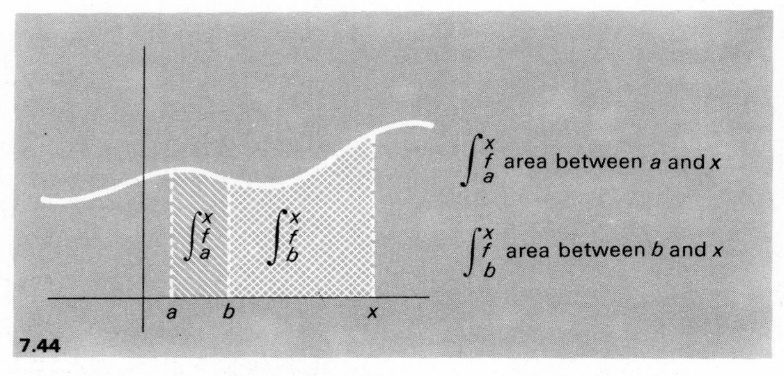

7.44

where, for the sake of argument, b is greater than a. The difference between the area $\int_a^x f$ and the area $\int_b^x f$ is simply the area between a and b: in other words $\int_a^b f$. This difference is *independent of* x, and so the arbitrariness in saying that $\int_0^x f = F(x)$ is only to the extent of a constant.

In fact we can be absolutely precise. We know that, however far we may be adrift in our assessment of what the function F is, the value of $\int_0^0 f$ is known: it must be zero. So, if it works out that the function we have chosen for F does not have $F(0) = 0$, we must adjust by writing

$$\int_0^x f = F(x) - F(0)$$

so that

$$\int_0^0 f = F(0) - F(0)$$

which is well and truly zero.

For example, we have seen that if $f(x) = 3x^2$, then we could well have chosen $F(x)$ as x^3 or $x^3 + 3$ or $x^3 + 10$, say. But whatever our choice of F, we still get the same answer for $F(x) - F(0)$: we get

$$\int_0^x f = x^3.$$

(If you cannot see this immediately, all you have to realize is that when you put $x = 0$, x^3 becomes zero, $x^3 + 3$ becomes 3 and $x^3 + 10$ becomes 10 – so in each case $F(x) - F(0)$ reduces to x^3.)

If you have managed, perhaps with several readings, to get a feel for what is going on, then you have made genuine progress towards understanding the differential and integral calculus, and there can be little doubt that the calculus is one of the towering achievements of the human intellect.

Of course there are a great many techniques, applications, new ideas, and difficulties to overcome if you want to achieve a working facility with the subject, but there are plenty of specialized texts available to help you.

If you are reading this present sentence then it may be the case that, unless you have the habit of finding out in advance if the butler did it, you have actually worked through to this point all the way from page 1 of the book. If you have, congratulations! Flick back over the pages and see how far you have come. You have seen how the use of symbols plays an essential part in mathematical work, you have met the idea of Cartesian geometry, you have grappled with the notion of a mathematical function, you have tried to unravel the mysteries of the concept of 'limit', and now you have made a preliminary attack on the differential and integral calculus. All these ideas are absolutely essential to modern mathematical thought and to have achieved even a partial insight into them is to have become party to one of the crowning glories of civilized thought. The greatest compliment that you could possibly pay the author is to resolve to take any one of the ideas you have met in this book just a little further by dipping into a more specialized text.

Index